THIS IS SEAT

teNeues

THE
CONTE

NTS

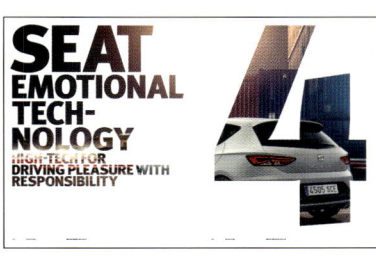

BRAND AND DESIGN BOOK

FEEL
THE

SUN

EN THE SUN IN THE SKY ABOVE BARCELONA IS AN INTENSE AND POWERFUL FORCE. IT BRINGS ENERGY AND INSPIRATION TO A CITY OF UNIQUE CREATIVITY.

ES EL SOL QUE ILUMINA BARCELONA ES UNA FUERZA INTENSA QUE LLENA DE ENERGÍA E INSPIRACIÓN A UNA CIUDAD CREATIVA COMO POCAS.

DE INTENSIV UND KRAFTVOLL STEHT DIE SONNE ÜBER BARCELONA. SIE LIEFERT ENER-GIE UND INSPIRATION FÜR EINE STADT MIT EINZIGARTIGER KREATIVITÄT.

TASTE THE

THIS IS SEAT   BRAND AND DESIGN BOOK

SEA

EN THE SEA THAT LIES BEFORE BARCELONA IS ONE OF DEPTH AND CLARITY. WATER IS LIFE, AND THIS CITY PACKED WITH YOUTHFUL EXUBERANCE SEEMS ALMOST TO DIVE RIGHT IN.

ES EL MAR QUE MECE BARCELONA REZUMA PROFUNDIDAD Y TRANSPARENCIA. EL AGUA ES VIDA, Y ESTA CIUDAD LLENA DE EXUBERANCIA JUVENIL CASI PARECE SUMERGIRSE EN ÉL.

DE KLARHEIT UND TIEFE BESITZT DAS MEER VOR BARCELONA. WASSER IST LEBEN, UND DIESE STADT VOLL VON JUGENDLICHER LEBENSFREUDE TAUCHT FAST EIN IN DIE SEE.

LOOK AT THE

SKY

EN THIS CITY BASKS IN A UNIQUE SHADE OF BLUE – A LIGHT THAT BRINGS FORMS TO LIFE AND MAKES THE PRECISION OF LINES TANGIBLE. IT IS A LIGHT THAT INSPIRES DESIGN.

ES LA CIUDAD, BAÑADA EN UN COLOR AZUL ÚNICO, TIENE UNA LUZ QUE DA VIDA A LAS FORMAS Y HACE TANGIBLE LA PRECISIÓN DE LAS LÍNEAS, UNA LUZ QUE INSPIRA DISEÑO.

DE EIN EINZIGARTIGES BLAU LÄSST DIESE STADT LEUCHTEN – EIN LICHT, DAS FORMEN LEBENDIG MACHT, DAS DIE PRÄZISION DER LINIEN SPÜRBAR MACHT. EIN LICHT, DAS DESIGN INSPIRIERT.

BREATHE

CITY

EN THIS IS A CITY YOU HAVE TO FEEL. IT IS THE VERY ESSENCE OF LIFE – YOUNG, DYNAMIC, CREATIVE. BARCELONA IS THE HOME OF AVANT-GARDE. BARCELONA IS THE HOME OF SEAT.

ES BARCELONA ES UNA CIUDAD PARA SENTIRLA. ES LA PURA ESENCIA DE LA VIDA: JOVEN, DINÁMICA, CREATIVA. BARCELONA ES SINÓNIMO DE VANGUARDIA. BARCELONA ES EL HOGAR DE SEAT.

DE DIESE STADT MUSS MAN SPÜREN. SIE IST LEBEN PUR, SIE IST JUNG, DYNAMISCH, KREATIV. BARCELONA IST DIE HEIMAT DER AVANTGARDE. BARCELONA IST DIE HEIMAT VON SEAT.

DRIVE THE

F U

UTURE

EN THIS IS AN AUTOMOBILE OF THE KIND THAT COULD ONLY HAVE BEEN CREATED IN BARCELONA – BY A DYNAMIC BRAND ON THE MOVE AND WITH A DESIGN THAT COMBINES PASSION WITH PRECISION.

ES ESTE ES UN COCHE QUE SOLO PODRÍA HABERSE CREADO EN BARCELONA, DE LA MANO DE UNA MARCA DINÁMICA EN MOVIMIENTO Y CON UN DISEÑO QUE COMBINA PASIÓN Y PRECISIÓN.

DE EIN AUTOMOBIL, WIE ES NUR IN BARCELONA ENTSTEHEN KONNTE. VON EINER MARKE, DIE IN DYNAMISCHER BEWEGUNG IST. MIT EINEM DESIGN, DAS PASSION MIT PRÄZISION VERBINDET.

SEAT IS ...
A BRAND ON THE MOVE

BRAND AND DESIGN BOOK

WITH A CLEAR VISION

EN THE MOST ASPIRATIONAL BRAND FOR YOUNG-SPIRITED, URBAN LIFESTYLE CUSTOMERS.

ES LA MARCA MÁS ATRACTIVA PARA CLIENTES DE ESPÍRITU JOVEN Y COSMOPOLITA.

DE DIE BEGEHRENSWERTESTE MARKE FÜR KUNDEN MIT JUNGEM UND URBANEM LEBENSGEFÜHL.

SEAT IS ...
TENSION

AL

EN Extreme tension focused entirely on that brief moment before the start. Every muscle is ready to deliver a perfect performance. That glorious second of sheer power – the 20V20 Concept Car from SEAT displays this power with every single detail of its design language. It is an athlete on mighty wheels, the vision of an enormously sporty and powerful SUV.

ES Extrema tensión justo antes de tomar la salida. Cada músculo está listo para funcionar a la perfección. Ese glorioso segundo de pura potencia – el showcar 20V20 de SEAT muestra ese poder en cada uno de los detalles de su lenguaje de diseño. Es un atleta de poderosos músculos, la visión de un SUV extremadamente deportivo y potente.

DE Höchste Spannung, absolut auf den Punkt. Der kurze Moment vor dem Start, jeder Muskel ist bereit zu perfekter Leistung. Die wunderbare Sekunde purer Kraft. Das 20V20 Concept Car von SEAT zeigt diese Kraft, mit jedem Detail seiner Designsprache. Es ist ein Athlet auf mächtigen Rädern, die Vision von einem sehr sportlichen und kraftvollen SUV.

SEAT IS ...
SCULPTU

RAL

EN A form full of life; a sculpture on wheels, expressed with just the right balance of light and shade, developed under the vibrant skies of Barcelona. The fine harmony of lines and surfaces and of power and elegance. The pure passion for movement, the joy of performance reflected in every detail. The 20V20 Concept Car shows the passion and the spirit of the SEAT brand – the pursuit of absolute precision of the kind that can only be achieved with exceptional quality and attention to detail.

ES Una forma llena de vida; una escultura sobre ruedas que se expresa con el equilibrio justo entre luces y sombras, desarrollada bajo los vibrantes cielos de Barcelona. Una delicada armonía de líneas y superficies, de potencia y elegancia. Pura pasión por el movimiento y alegría de prestaciones que se reflejan en cada uno de sus detalles. El showcar 20V20 muestra la pasión y el espíritu de la marca SEAT: una búsqueda incansable de la precisión absoluta, aquella que solo puede alcanzarse con una calidad excepcional y una minuciosa atención a cada detalle.

DE Eine Form voller Leben, eine Skulptur auf Rädern. Die richtige Balance von Licht und Schatten, entwickelt unter dem strahlenden Himmel von Barcelona. Die feine Harmonie von Linien und Flächen, von Kraft und Eleganz. Die pure Lust an der Bewegung, die Freude an der Dynamik spiegeln sich in jedem Detail. Das 20V20 Concept Car zeigt die Passion und den Spirit der Marke SEAT – den Willen zu absoluter Präzision, wie sie nur mit höchster Qualität und Liebe zum Detail entstehen kann.

SEAT IS ...
CHARACT

TERFUL

EN A purposeful gaze, focused on a point in the distance. A clear direction, always forward. An unmistakeable expression, determined and powerful. The 20V20 Concept Car from SEAT is fully concentrated on the essentials, displaying the clear character of the Spanish brand in a whole new dimension.

ES Una mirada llena de intención, enfocada hacia un punto lejano. Una dirección clara: siempre hacia delante. Una expresión inconfundible, con inteligencia, determinación y potencia. El showcar 20V20 de SEAT está totalmente centrado en lo esencial, mostrando el carácter claro de la marca española en una dimensión completamente nueva.

DE Ein klarer Blick, fokussiert auf einen Punkt in der Ferne. Eine klare Richtung, immer nach vorne. Ein unverwechselbares Gesicht, entschlossen und kraftvoll. Das 20V20 Concept Car von SEAT besitzt diese Konzentration auf das Wesentliche, es zeigt den klaren Charakter der spanischen Marke in einer neuen Dimension.

SEAT △ LIGHTING

LUCA DE MEO'S PERSPECTIVE

"ENGINEERED BEAUTY" EXECUTED WITH ABSOLUTE CONSISTENCY. THE BARCELONA PAVILION BY MIES VAN DER ROHE PROVIDES INSPIRATION FOR THE DEVELOPMENT OF A STRONG BRAND.

ES «Engineered beauty» ejecutada con absoluta solidez. El pabellón de Barcelona, de Mies van der Rohe, es fuente de inspiración para el desarrollo de una marca fuerte.
DE „Engineered beauty" in absoluter Konsequenz.
Der Barcelona-Pavillon von Mies van der Rohe liefert Inspiration für die Entwicklung einer starken Marke.

EN Luca de Meo,
Chairman of the Executive Committee, SEAT S.A.
ES Luca de Meo,
Presidente del Comité Ejecutivo de SEAT S.A.
DE Luca de Meo,
Vorsitzender des Vorstands der SEAT S.A.

Outside and inside meld into a single space; despite being heavy materials, the stone and steel generate a pervading sense of lightness. When he created the highly innovative Barcelona Pavilion in 1928, German architect Ludwig Mies van der Rohe built a milestone of modern architecture.

ES
El exterior y el interior se funden en un solo espacio; a pesar de los materiales pesados, la piedra y el acero transmiten una penetrante sensación de levedad. Cuando el arquitecto alemán Ludwig Mies van der Rohe creó en 1928 el innovador pabellón de Barcelona, marcó un hito en la historia de la arquitectura moderna.

DE
Außen und Innen verschmelzen zu einem Raum, die schweren Materialien Stein und Stahl schaffen ein Gefühl vollkommener Leichtigkeit. Mit dem extrem innovativen Barcelona-Pavillon hat der deutsche Architekt Ludwig Mies van der Rohe 1928 einen Meilenstein der modernen Architektur gebaut.

ENDURING BEAUTY AND QUALITY CAN ONLY BE CREATED WITH A CLEAR IDEA IN MIND.

ES Belleza y calidad duraderas solo se pueden crear con una idea clara en la cabeza.
DE Dauerhafte Schönheit und Qualität entstehen nur mit einer klaren Idee im Kopf.

EN   It is the consistency that Luca de Meo finds so fascinating about this construction – the simplicity and logic, the absolute concept. One of the most influential buildings in the history of modern architecture was created here in the middle of Barcelona almost 90 years ago. And it remains compelling to this day, for its absolute minimalism, the coherence of its proportions and the quality of the materials. Nothing was left to chance; the tiniest detail is part of the bigger plan. For the new boss, this is "engineered beauty"; precisely crafted, enduring beauty.

Luca de Meo knows this place well. He loves the Barcelona Pavilion, designed by Ludwig Mies van der Rohe in 1928 for the World Exhibition in the Catalan city. 1928! Back then, such modernity was a level of exoticism barely imaginable these days. In this exuberant phase of history, the pavilion stood in absolute contrast to its surroundings. And yet, in its clarity and focus, this building in particular possesses something unsurpassed, something utterly timeless, even in the year 2015.

IN ITS CLARITY, THIS BUILDING POSSESSES SOMETHING UTTERLY TIMELESS, EVEN IN THE YEAR 2015.

For the man at the head of the Spanish company, this building is a place of inspiration – and of parallels. A strong brand needs characteristics that Mies van der Rohe gave his pavilion. Consistency in its idea and implementation, clarity in all dimensions and quality of execution are at the very top of the list for Luca de Meo. Absolute openness to new ideas is just as much a part of this as the enormous dynamism of its execution. For the SEAT brand, an important step towards this new power is the design language and quality of the Leon, in its premium simplicity and, at the same time, sculptural clarity.

The "Leon formula" points the way forward for the brand – emotionality in perfect combination with technology and quality; Spanish passion unites with German perfection. It is a union of two worlds that makes SEAT unique – and that Luca de Meo lives and breathes as passion as an individual. He is Italian, he loves passion, but also the precision he worked with during his years with Volkswagen and Audi in Germany.

For de Meo, the SEAT brand possesses enormous potential. The task over the next few years is to develop it and to implement it successfully. The aim and the strategy are clear; the implementation is being consistently pursued. For Luca de Meo, the SEAT 20V20 concept car embodies the vision of the brand a few years from now – here, too, with clarity that promises to endure.

ES   Es la rotundidad de esta construcción lo que fascina a Luca de Meo: la simplicidad y la lógica, el concepto de lo absoluto. Uno de los edificios más influyentes de la historia de la arquitectura moderna se creó aquí, en el centro de Barcelona, hace casi 90 años. Esta fascinación perdura hasta el día de hoy por su absoluto minimalismo, el equilibrio de sus proporciones y la calidad de los materiales. Nada se dejó al azar; el detalle más ínfimo forma parte de un plan mayor. El nuevo presidente de SEAT lo llama «engineered beauty» o ingeniería de la belleza: belleza duradera fabricada con absoluta precisión.

Luca de Meo conoce bien la ciudad. Siente pasión por el Pabellón de Barcelona diseñado por Ludwig Mies van der Rohe en 1928 para la Exposición Internacional de la capital catalana. ¡1928! En aquellos años esta modernidad supuso un exotismo casi inimaginable. En aquel momento exuberante de la historia, el pabellón resaltaba por el contraste absoluto con su entorno. Y, sin embargo, es en esa claridad y esa precisión donde reside la esencia de este edificio, que posee algo incomparable y absolutamente atemporal, todavía hoy en el año 2015.

ES EN LA SIMPLICIDAD DE ESTE EDIFICIO DONDE RESIDE SU ESENCIA ABSOLUTAMENTE ATEMPORAL, INCLUSO EN EL AÑO 2015.

Para el hombre que lleva las riendas de la empresa española, este edificio es un lugar de inspiración y de paralelismos. Una marca sólida necesita las características con las que Mies van der Rohe concibió este pabellón. Coherencia en la idea y en la puesta en práctica, claridad en todas las dimensiones y calidad de ejecución son las prioridades de Luca de Meo. La apertura total a ideas nuevas es tan importante como el enorme dinamismo de su ejecución. Para la marca SEAT, un importante paso adelante hacia ese nuevo poder es el lenguaje del diseño y la calidad del León, por su simplicidad superior y, al mismo tiempo, por la claridad escultural.

La «fórmula León» es el camino que debe seguir la marca: emoción en perfecta armonía con tecnología y calidad, la pasión española unida a la perfección alemana. Es la combinación de los dos mundos lo que hace de SEAT una marca única y Luca de Meo cree firmemente en ello. De origen italiano, es amante de la pasión y de la precisión con la que ha trabajado durante años para Volkswagen y Audi en Alemania.

Luca de Meo afirma que la marca SEAT posee un enorme potencial. La tarea de estos próximos años es desarrollarla e implementarla con éxito. El objetivo y la estrategia están claros; la implementación se debe realizar de forma constante. Para de Meo, el showcar 20V20 de SEAT representa la visión de la marca de aquí a unos años, con una claridad que promete perdurar.

DE   Es ist die Konsequenz, die Luca de Meo an diesem Bauwerk fasziniert. Die Einfachheit und Logik, das absolute Konzept. Vor fast 90 Jahren ist hier mitten in Barcelona eines der einflussreichsten Gebäude der modernen Architekturgeschichte entstanden. Und es begeistert noch heute durch seine absolute Reduktion, die Stimmigkeit der Proportionen, die Qualität der Materialien. Hier wurde nichts dem Zufall überlassen, hier ist das kleinste Detail Teil des großen Plans. „Engineered beauty" ist das für den neuen Chef von SEAT, präzise gestaltete, dauerhafte Schönheit.

Luca de Meo kennt diesem Platz gut. Er schätzt den Barcelona-Pavillon, den Ludwig Mies van der Rohe 1928 für die Weltausstellung in der katalanischen Metropole entworfen hat. 1928! Die Moderne hatte damals etwas heute kaum vorstellbar Exotisches, der Pavillon stand in dieser überschäumenden Phase der Geschichte in maximalem Kontrast zu seiner Umgebung. Und doch besitzt gerade dieses Gebäude in seiner Klarheit, in seiner Fokussiertheit auf das Wesentliche auch noch im Jahr 2015 etwas nicht zu Steigerndes, vollkommen Zeitloses.

IN SEINER KLARHEIT BESITZT DIESES BAUWERK AUCH IM JAHR 2015 ETWAS VOLLKOMMEN ZEITLOSES.

Für den Mann an der Spitze des spanischen Unternehmens ist dieses Gebäude ein Ort der Inspiration. Und der Parallelen. Eine starke Marke braucht Eigenschaften, wie sie Mies van der Rohe seinem Pavillon mitgegeben hat. Konsequenz in Idee und Umsetzung, Klarheit in allen Dimensionen und Qualität in der Ausführung stehen für Luca de Meo dabei ganz oben. Die absolute Offenheit gegenüber neuen Ideen gehört ebenso dazu wie die enorme Dynamik, mit der sie umgesetzt werden. Ein wichtiger Schritt der Marke SEAT hin zu dieser neuen Kraft ist die Designsprache und die Qualität des Leon in ihrer hochwertigen Einfachheit und zugleich skulpturhaften Klarheit.

Die „Leon-Formel" beschreibt den Weg der Marke in die Zukunft: Emotionalität in perfekter Kombination mit Technologie und Qualität, die spanische Passion vereint mit der deutschen Perfektion. Eine Verbindung von zwei Welten, die SEAT einzigartig macht – und die Luca de Meo auch als Person lebt: Er ist Italiener, er liebt die Leidenschaft, aber ebenso die Präzision, mit der er schon in seinen Jahren bei Volkswagen und Audi in Deutschland gearbeitet hat.

Die Marke SEAT hat für de Meo enormes Potenzial, es zu fördern und in Erfolg umzusetzen, ist die Aufgabe der nächsten Jahre. Das Ziel und die Strategie sind klar, die Umsetzung geschieht mit Konsequenz. Das SEAT 20V20 Concept Car verkörpert für Luca de Meo die Vision der Marke in ein paar Jahren – auch hier mit einer Klarheit, die Dauerhaftigkeit verspricht.

EN
Marble, onyx and travertine give solidity, while large expanses of glass and fine steel pillars generate the feeling of lightness, before the reflections on the water ultimately cause the spaces to disperse. The Barcelona Chair by Mies van der Rohe also became an enduring classic.

ES
El mármol, el ónice y el travertino imprimen solidez, y las grandes superficies de cristal y las elegantes columnas de acero producen una sensación de ligereza, con un espacio que queda disipado por el reflejo en el agua. La silla Barcelona de Mies van der Rohe también se convirtió en un clásico perdurable.

DE
Marmor, Onyx und Travertin geben Solidität, große Glasflächen und feine Stahlstützen erzeugen das Gefühl von Leichtigkeit. Und die Spiegelungen des Wassers lösen die Räume endgültig auf. Zu einem dauerhaften Klassiker wurde auch van der Rohes Barcelona-Chair.

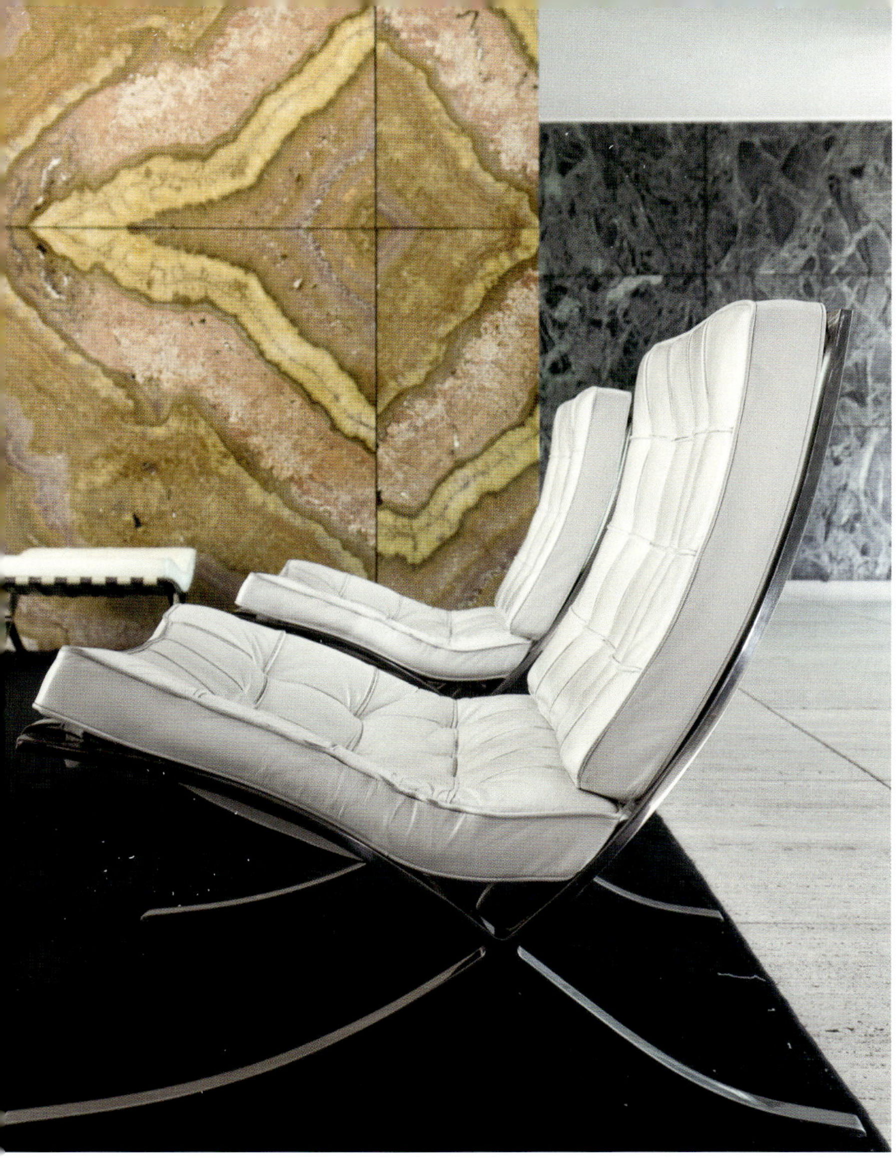

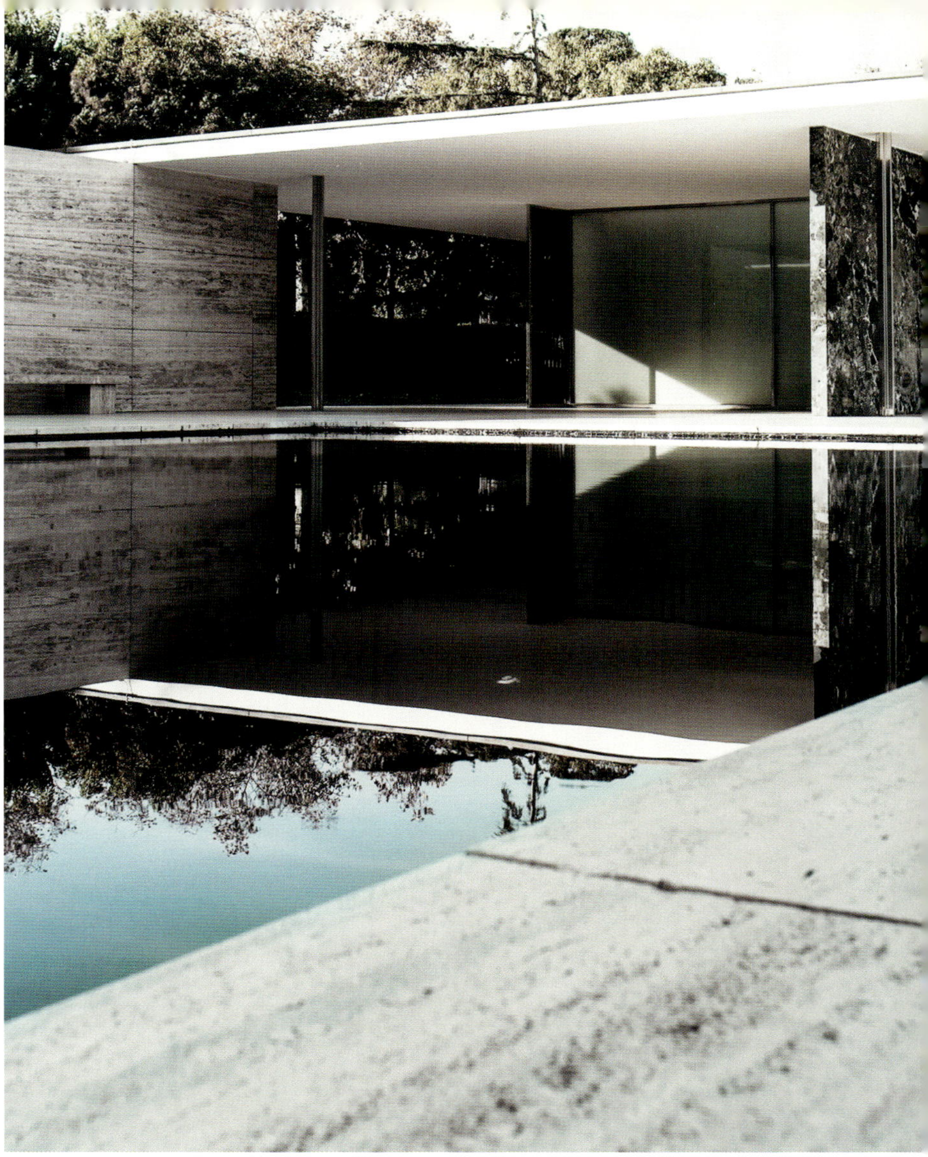

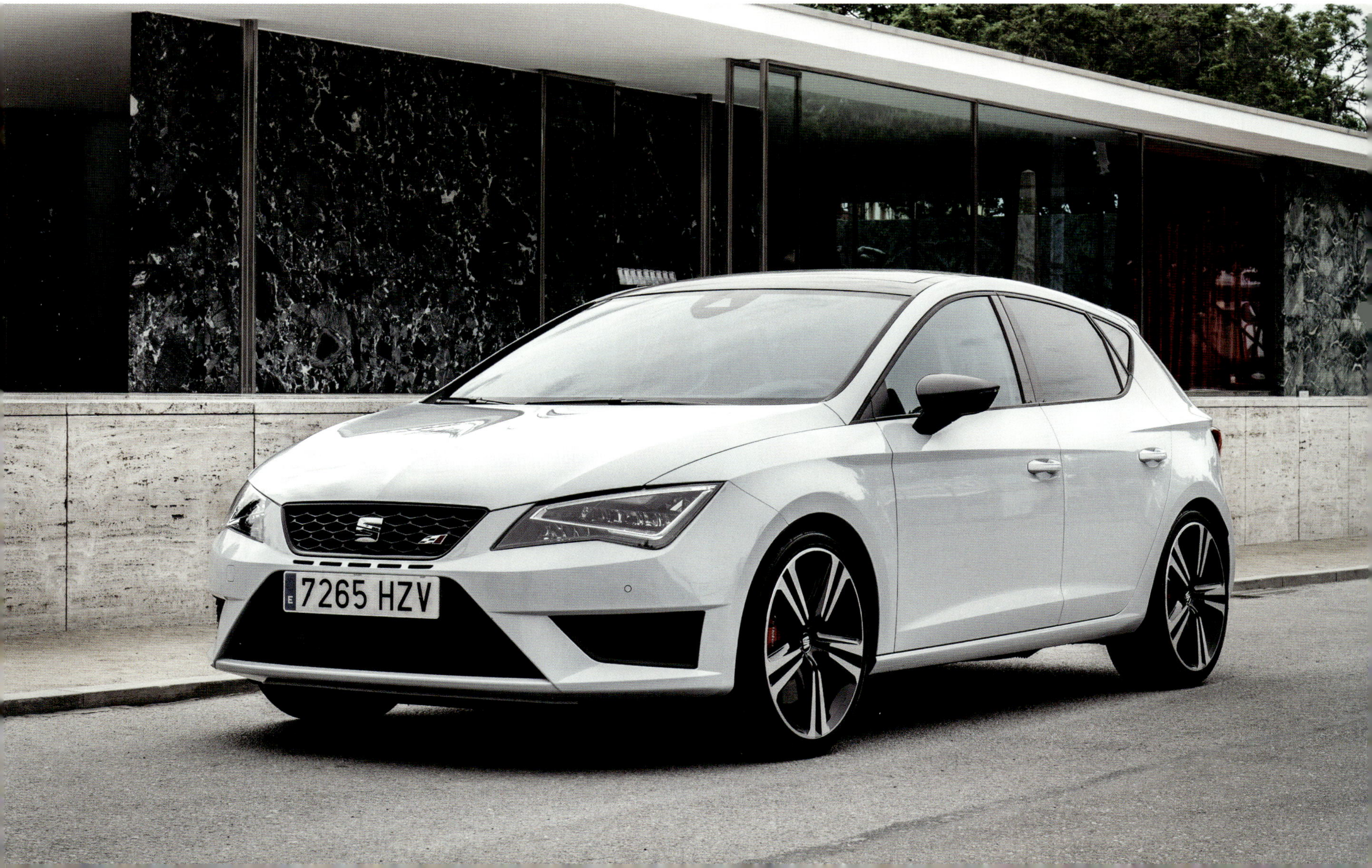
7265 HZV

PART OF MY LIFE

TRAVEL IN STYLE
A DAY IN BARCELONA

EN Barcelona possesses the very particular magic of a vibrant city by the sea. The warm light of early morning mingles with the freshness of the water. The day begins – a new day in the youthful life of the city.

ES Barcelona posee la magia particular de una ciudad vibrante junto al mar. La luz cálida de la mañana se mezcla con el frescor del agua. Empieza un nuevo día en la vida juvenil de la ciudad.

DE Barcelona besitzt den ganz besonderen Zauber einer lebendigen Stadt am Meer. Das warme Licht des frühen Morgens verbindet sich mit der Frische des Wassers. Der Tag beginnt. Ein neuer Tag im jungen Leben der Metropole.

EN The road into the city leads to Torre Agbar – 40 colours reflect the sunlight, 4,500 adjustable glass louvres regulate the energy balance. Architecture and design with sustainable function. Torre Agbar is a symbol of Barcelona and of a modern, urban lifestyle.

ES Una de las arterias de la ciudad pasa por la Torre Agbar: 40 colores reflejan la luz del sol, 4.500 lamas de vidrio ajustables regulan el equilibrio de energía. Arquitectura y diseño con una función sostenible. La Torre Agbar es un símbolo de Barcelona y la expresión de un estilo de vida moderno y urbano.

DE Der Weg in die Stadt führt zum Torre Agbar. 40 Farben reflektieren das Sonnenlicht, 4.500 Glaslamellen regulieren den Energiehaushalt. Architektur und Design mit nachhaltiger Funktion. Der Torre Agbar ist ein Symbol für Barcelona. Und für einen modernen, urbanen Lebensstil.

INNOVATION

EN The first coffee of the day beneath the palm trees along the promenade. A shot of energy and inspiration for the day ahead – accompanied by a car with its own, unique charisma.

ES El primer café del día bajo las palmeras del paseo marítimo. Una inyección de energía e inspiración para el día que empieza, al volante de un coche con un carisma propio y único.

DE Ein erster Kaffee unter den Palmen der Promenade. Energie und Inspiration für den Tag – begleitet von einem Automobil, das eine ganz eigene Ausstrahlung besitzt.

INSPIRATION

YOUNG

53

EN The Mediterranean light seems to concentrate itself in this colour.
Exuding extreme brilliance, with an impressive sense of depth. SEAT is energy!

ES La luz del Mediterráneo parece cobrar intensidad con este color. Exuda un
brillo propio con una sensación impresionante de profundidad. SEAT es energía.

DE Das mediterrane Licht scheint sich in dieser Farbe zu konzentrieren.
Extreme Leuchtkraft, beeindruckende Tiefenwirkung, Kraft pur. SEAT ist Energie!

ENERGY

URBAN

EN Barcelona is cosmopolitan – with myriad impressions and influences, an utterly boundless diversity of possibilities. Communication shapes urban life – contact with your neighbour is just as important as the connection to friends on the other side of the world. Urban lifestyle is not bound to the city, it is a mind-set – a love of life!

ES Barcelona es cosmopolita, rebosante de impresiones e influencias y una diversidad infinita de posibilidades. La comunicación moldea la vida urbana: el contacto con los vecinos es tan importante como la relación con los amigos en la otra parte del mundo. La vida urbana no se limita a la ciudad, es más bien una actitud vital: disfrutar de la vida.

DE Barcelona bedeutet Urbanität – zahllose Eindrücke und Einflüsse, eine schier grenzenlose Vielfalt an Möglichkeiten. Kommunikation prägt das urbane Leben, der Kontakt zum Nachbarn zählt ebenso wie die Verbindung zu Freunden am anderen Ende der Welt. Urbaner Lebensstil ist nicht an die Stadt gebunden, es ist eine Haltung: das Leben genießen!

EN Every SEAT is a car in which you feel safe, well protected, yet closely connected with the world. It is a car like a good friend – strong and attentive.

ES Todo SEAT es un coche en el que te sientes seguro, bien protegido pero, a la vez, bien conectado con el mundo. Es como un buen amigo: sólido y atento.

DE Jeder SEAT ist ein Automobil, in dem man sich geborgen fühlt, gut beschützt und doch mit der Welt eng verbunden. Ein Auto wie ein guter Freund – kraftvoll und aufmerksam.

EN **An active and youthful lifestyle is a good match for a SEAT – cosmopolitan, yet solid, dynamic and versatile.**

ES Un estilo de vida activo y juvenil hace buena pareja con un SEAT: cosmopolita pero también sólido, dinámico y versátil.

DE Zu einem SEAT passt ein aktiver und junger Lebensstil, weltoffen und doch solide, dynamisch und vielseitig.

OPEN MINDED

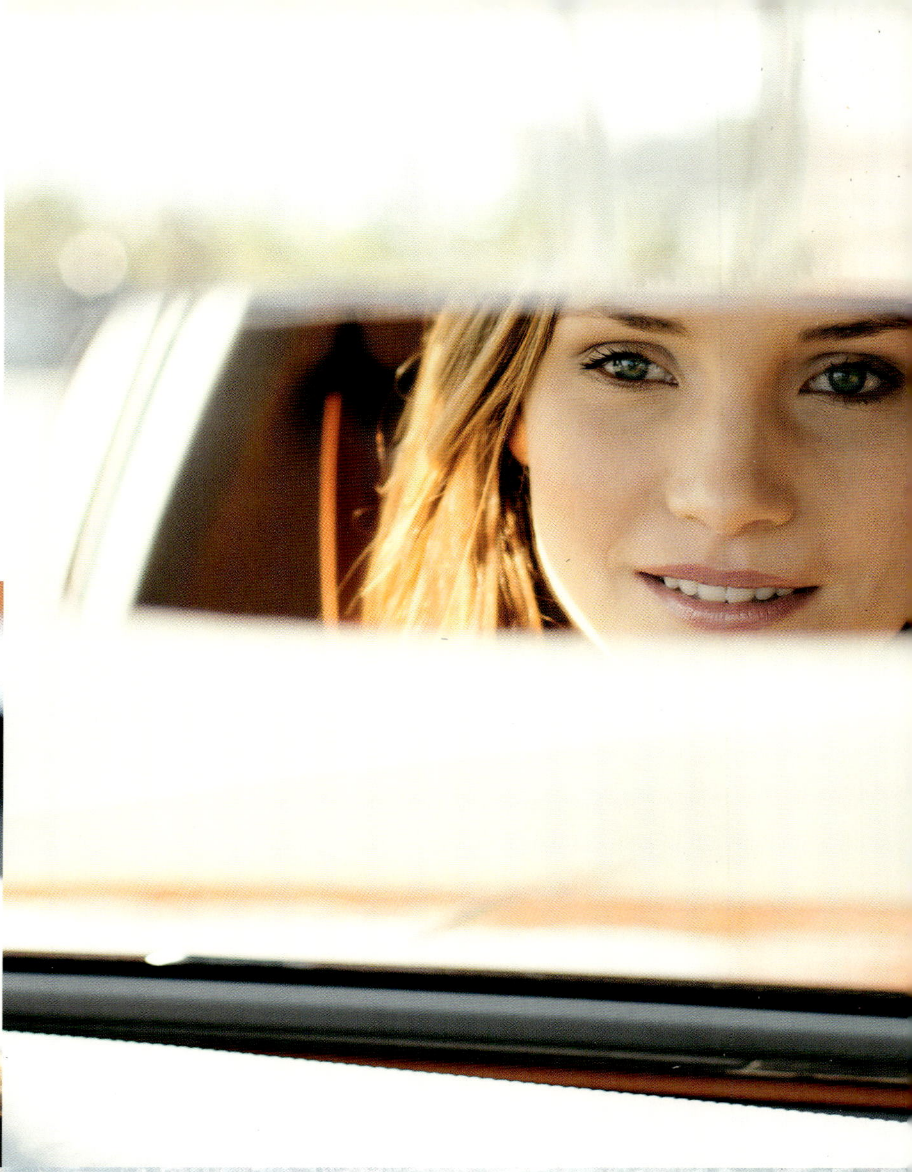

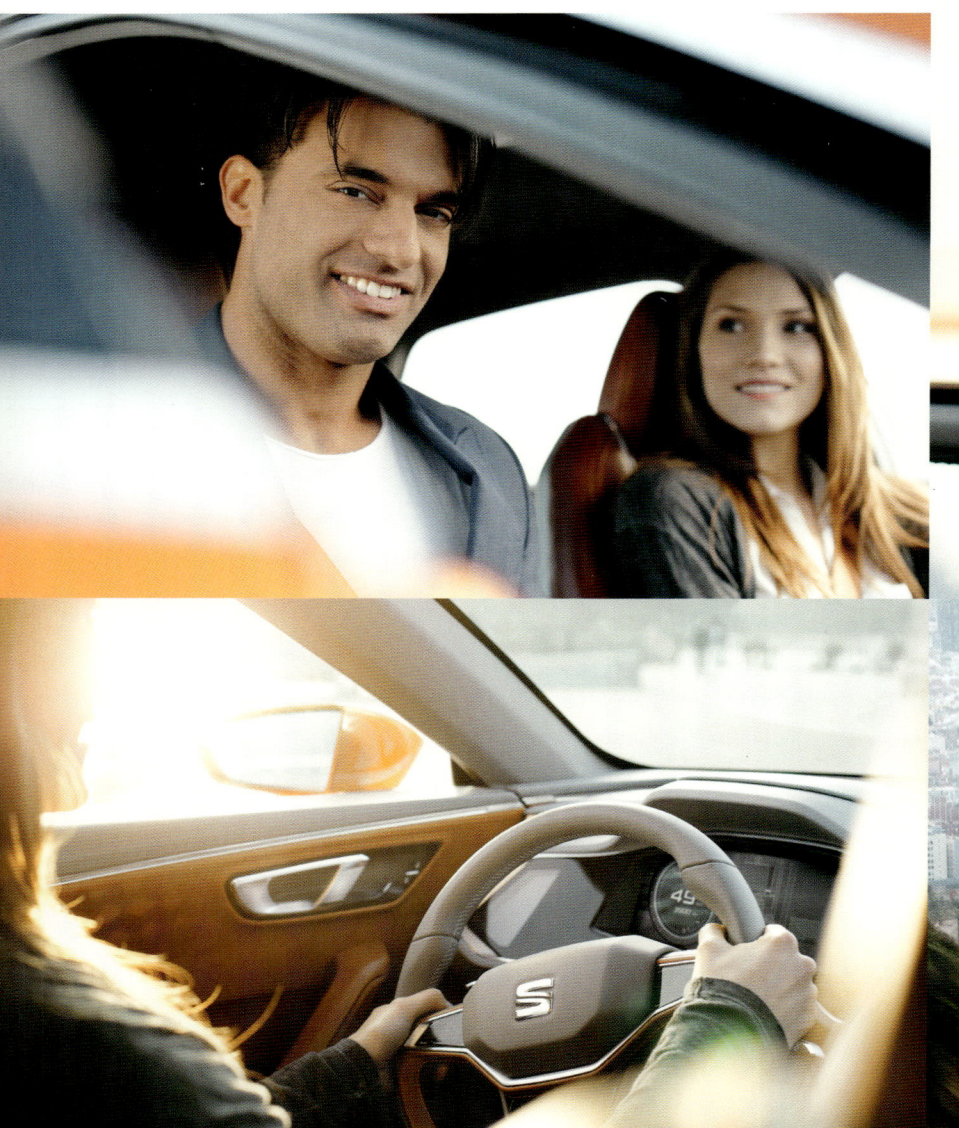

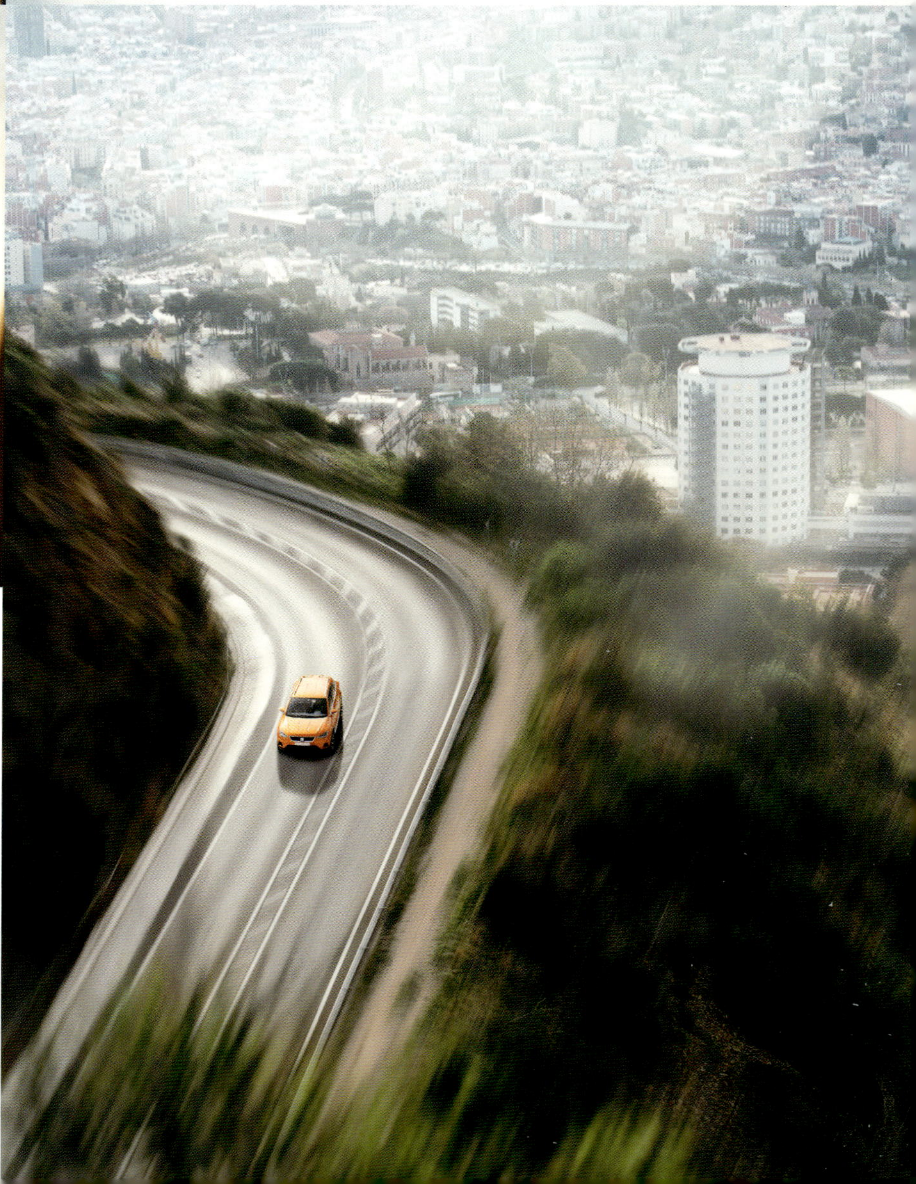

EN As the evening approaches, the colours become more dramatic – and sometimes life does, too. Barcelona is a city in motion – 24 hours a day. SEAT is a reliable companion.

ES Al atardecer, los colores se vuelven más interesantes, y a veces la vida también. Barcelona es una ciudad en movimiento las 24 horas del día. SEAT es un compañero fiable.

DE In den Abendstunden gewinnen die Farben an Dramatik und manchmal auch das Leben. Barcelona ist in Bewegung, 24 Stunden am Tag. SEAT ist ein verlässlicher Begleiter.

LIFE

SEAT
WHO
WE ARE

DESIGN

ES DISEÑO
DE DESIGN

DYNAMICS

ES DINÁMICA
DE DYNAMIK

CUSTOMER SATISFACTION

ES SATISFACCIÓN DE LOS CLIENTES
DE KUNDEN-ZUFRIEDENHEIT

CUSTOMER

ES CLIENTE
DE KUNDE

INDIVIDUAL PERSONALITY

ES PERSONALIDAD INDIVIDUAL
DE INDIVIDUELLE PERSÖNLICHKEIT

STATE-OF-THE-ART TECHNOLOGY

ES TECNOLOGÍA DE VANGUARDIA
DE STATE-OF-THE-ART-TECHNOLOGIE

MORE THAN 14,000 EMPLOYEES

ES MÁS DE 14.000 EMPLEADOS
DE MEHR ALS 14.000 MITARBEITER

THE BRAND

FORMULA

DESIGN & FUNCTIONALITY
DYNAMISM & COMFORT
ACCESSIBILITY & HIGHEST QUALITY
EMOTION & TECHNOLOGY

YOUNG SPIRIT

QUALITY

ES CALIDAD
DE QUALITÄT

ES

LA FÓRMULA
DE ÉXITO DE LA MARCA:

DISEÑO Y FUNCIONALIDAD
DINAMISMO Y CONFORT
ACCESIBILIDAD Y CALIDAD SUPERIOR
EMOCIÓN Y TECNOLOGÍA

ESPÍRITU JOVEN

DE

DIE MARKEN-
FORMEL:

DESIGN & FUNKTIONALITÄT
DYNAMIK & KOMFORT
ERREICHBARKEIT & HÖCHSTE QUALITÄT
EMOTION & TECHNOLOGIE

JUNGER SPIRIT

SEAT CONNECTS THE BEST OF TWO WORLDS!

SPAIN
DESIGN
PASSION
QUALITY OF LIFE

ENGINEERS
OF BEAUTY

GERMANY
PERFORMANCE
QUALITY
EFFICIENCY

PASSIONATE
PERFECTIONISTS

ES **ESPAÑA**
DISEÑO
PASIÓN
CALIDAD DE VIDA

ENGINEERS
OF BEAUTY

DE **SPANIEN**
DESIGN
PASSION
LEBENSQUALITÄT

ENGINEERS
OF BEAUTY

ES **ALEMANIA**
RENDIMIENTO
CALIDAD
EFICIENCIA

PASSIONATE
PERFECTIONISTS

DE **DEUTSCHLAND**
LEISTUNG
QUALITÄT
EFFIZIENZ

PASSIONATE
PERFECTIONISTS

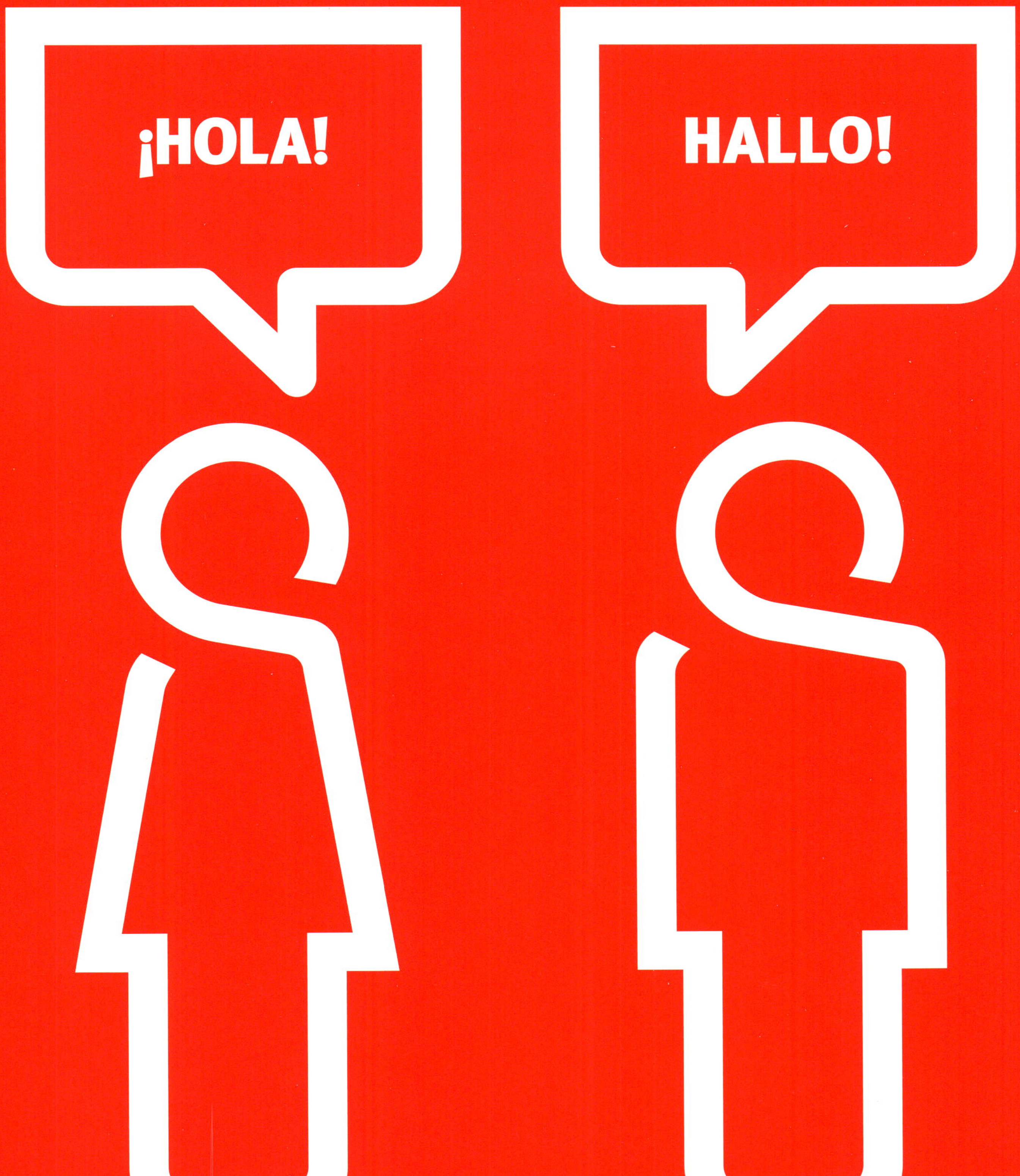

WE ARE SPANISH

WE ARE GERMAN

THE ONLY TRULY SPANISH CAR MANUFACTURER

THE SEAT BRAND STANDS FOR PASSION AND THE LOVE OF CARS, FOR FASCINATING DESIGN AND OUTSTANDING TECHNICAL EXPERTISE. SEAT IS THE ONLY CAR MAKER TO DESIGN, DEVELOP, PRODUCE AND MARKET ITS MODELS ENTIRELY IN SPAIN.

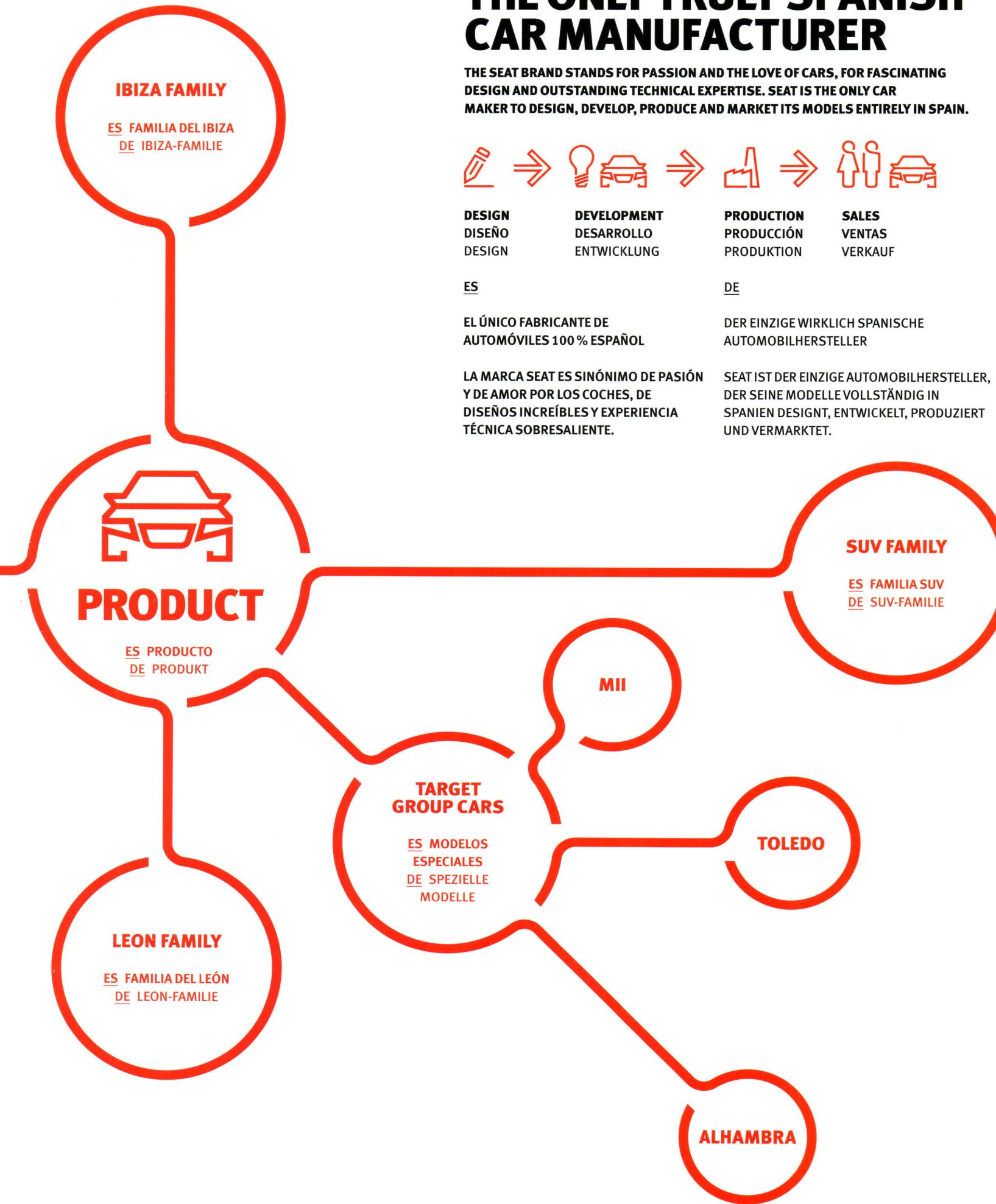

DESIGN
DISEÑO
DESIGN

DEVELOPMENT
DESARROLLO
ENTWICKLUNG

PRODUCTION
PRODUCCIÓN
PRODUKTION

SALES
VENTAS
VERKAUF

ES

EL ÚNICO FABRICANTE DE
AUTOMÓVILES 100 % ESPAÑOL

LA MARCA SEAT ES SINÓNIMO DE PASIÓN
Y DE AMOR POR LOS COCHES, DE
DISEÑOS INCREÍBLES Y EXPERIENCIA
TÉCNICA SOBRESALIENTE.

DE

DER EINZIGE WIRKLICH SPANISCHE
AUTOMOBILHERSTELLER

SEAT IST DER EINZIGE AUTOMOBILHERSTELLER,
DER SEINE MODELLE VOLLSTÄNDIG IN
SPANIEN DESIGNT, ENTWICKELT, PRODUZIERT
UND VERMARKTET.

IBIZA FAMILY
ES FAMILIA DEL IBIZA
DE IBIZA-FAMILIE

PRODUCT
ES PRODUCTO
DE PRODUKT

SUV FAMILY
ES FAMILIA SUV
DE SUV-FAMILIE

MII

TARGET GROUP CARS
ES MODELOS ESPECIALES
DE SPEZIELLE MODELLE

TOLEDO

LEON FAMILY
ES FAMILIA DEL LEÓN
DE LEON-FAMILIE

ALHAMBRA

SUCCESS
ES ÉXITO
DE ERFOLG

GROWING SALES
ES CRECIMIENTO DE LAS VENTAS
DE WACHSENDER ABSATZ

SUSTAINABLE COMPANY
ES COMPAÑIA SOSTENIBLE
DE NACHHALTIGES UNTERNEHMEN

SEAT

TEAM
ES EQUIPO
DE TEAM

MOST ATTRACTIVE EMPLOYER
ES EMPLEADOR MÁS ATRACTIVO
DE ATTRAKTIVSTER ARBEITGEBER

DEALERSHIPS & SUPPLIERS WITH THOUSANDS OF MORE JOBS
ES CONCESIONARIOS Y PROVEEDORES CON MILES DE PUESTOS DE TRABAJO
DE HÄNDLER UND ZULIEFERER MIT TAUSENDEN WEITEREN JOBS

DUAL EDUCATION AND TRAINING
ES EDUCACIÓN DUAL Y FORMACIÓN
DE AUSBILDUNG UND TRAINING

BRAND
ICON
IBIZA

THE BRAND PILLARS

IBIZA
LEON
SUV

EN The Ibiza led SEAT out of Spain and into the rest of the world. For more than 30 years, the Ibiza has been the brand's most successful model, with well over five million Ibizas rolling off the production line in Barcelona.

In its latest generation, the Ibiza family is one of the most attractive models in the segment – stunning design, dynamic fun to drive, premium interior, efficient drive and leading connectivity guarantee enduring success.

With its extensive options for individualisation, the Ibiza has a high proportion of young and female customers. However, the compact model range is also becoming increasingly popular among older drivers, the so-called "empty nesters". They appreciate the Ibiza's practical format and light-footed agility.

The Ibiza will remain a major cornerstone of the brand going forward – with fascinating design and innovative concepts.

ES Con el Ibiza, SEAT traspasó la frontera española y se dio a conocer al resto del mundo. Durante más de 30 años, el Ibiza ha sido el modelo más exitoso de la marca, con más de cinco millones de unidades fabricadas en la línea de producción de Barcelona.

En su última generación, la familia del Ibiza es uno de los modelos más atractivos de su segmento: diseño espectacular, conducción dinámica y divertida, interior de calidad superior, conducción eficiente y conectividad avanzada que garantizan un éxito duradero.

Con una amplia oferta de opciones de personalización, el Ibiza es favorito de mujeres y jóvenes. Pero la gama compacta del modelo es cada vez más popular entre conductores más mayores, los conocedores del término «síndrome del nido vacío», que aprecian el formato práctico y la agilidad dinámica del Ibiza.

El Ibiza permanecerá como un referente clave de la marca en movimiento con un diseño fascinante y conceptos innovadores.

DE Der Ibiza hat SEAT aus Spanien in die Welt geführt: Seit mehr als 30 Jahren ist der Ibiza das erfolgreichste Modell der Marke, weit mehr als fünf Millionen Ibiza sind aus den Produktionshallen bei Barcelona gerollt.

In ihrer neuesten Generation zählt die Ibiza-Familie zu den attraktivsten Modellen in ihrem Segment: Begeisterndes Design, dynamischer Fahrspaß, hochwertiges Interieur, effizienter Antrieb und führende Connectivity garantieren den dauerhaften Erfolg.

Der Ibiza mit seinen vielfältigen Möglichkeiten zur Individualisierung hat einen hohen Anteil junger und weiblicher Kunden. Zunehmend wird die kompakte Modellfamilie aber auch interessant für ältere Fahrerinnen und Fahrer, die „Empty Nesters". Sie schätzen das handliche Format und die leichtfüßige Agilität des Ibiza.

Auch in Zukunft wird der Ibiza eine wesentliche Stütze der Marke bleiben – mit faszinierendem Design und innovativen Konzepten.

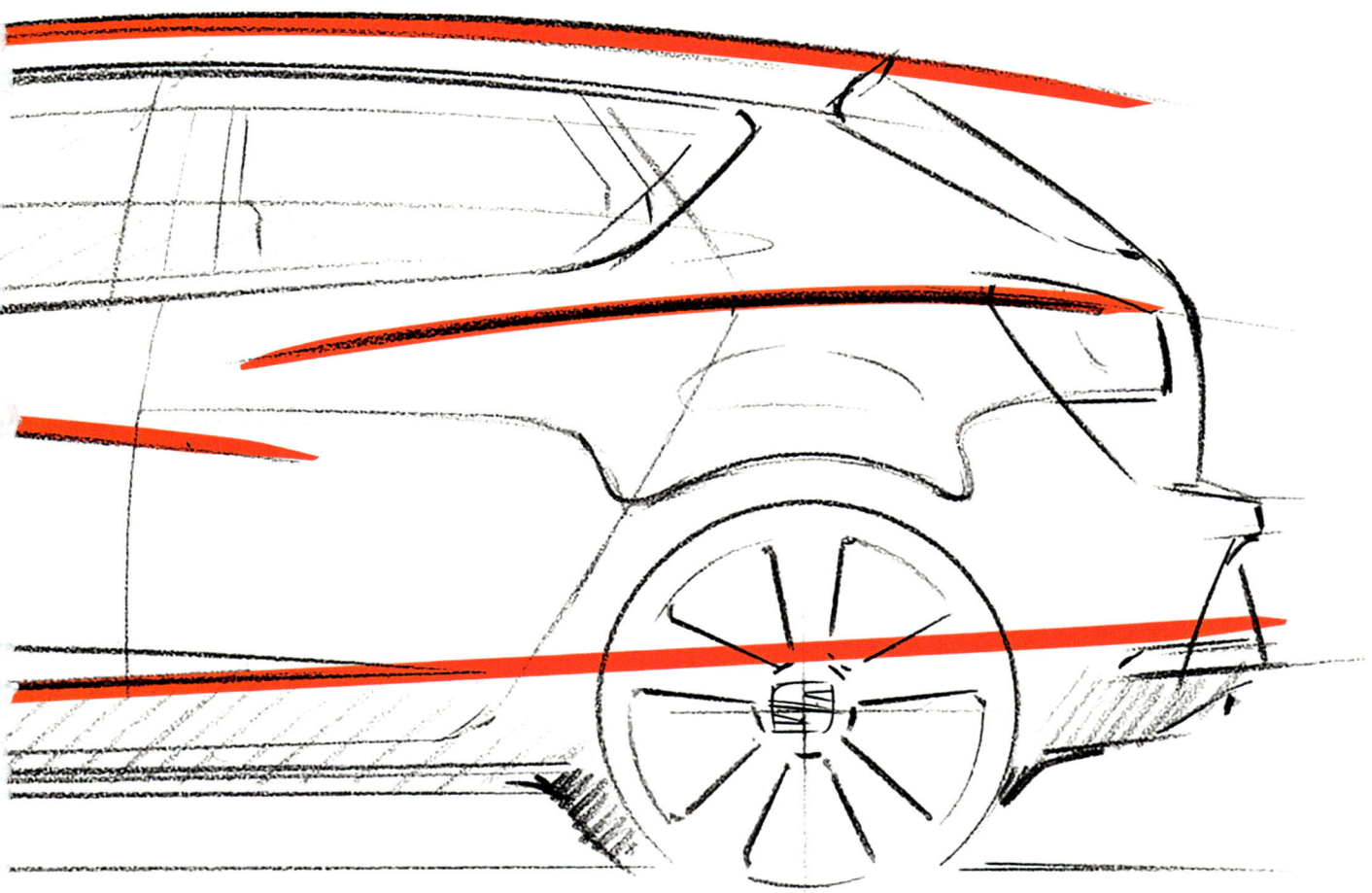

BRAND
ICON
LEON

THE BRAND PILLARS

IBIZA
LEON
SUV

EN The current generation of the Leon is extremely successful and is taking the entire SEAT brand to a whole new level. Customers are giving it top marks for design and quality, performance and utility, and technology and reliability. And the trade press is equally impressed, placing the Leon on the podium in around 75 percent of all magazine comparison tests across Europe.

The Leon family is now the brand's second biggest line. The diversity of the range means the Leon appeals to a broad target group – from the Leon Ecomotive, which pairs extreme efficiency with driving fun, to the Leon CUPRA, which pairs extreme driving fun with efficiency, or the Leon X-PERIENCE, which is virtually unlimited in its versatility. Design is always a major factor in the purchasing decision for the Leon target group, as are reliability, driving fun, economy and connectivity. The Leon's many qualities have made it a benchmark and role model for every product bearing the SEAT logo.

ES La generación actual del León goza de un éxito arrollador, que impulsa a SEAT a una nueva dimensión. Los clientes están dando a la marca notas muy altas en cuanto a diseño, calidad, comportamiento, uso, y tecnología y fiabilidad. La prensa especializada también está muy impresionada: un 75 por ciento de las comparativas de todas las revistas en Europa coloca al León en el podio.

La familia del León es ahora la segunda mayor línea de la marca. La diversidad de la gama ha conseguido que el León atraiga a un amplio grupo objetivo: desde el León Ecomotive, que combina eficiencia superior y diversión en la conducción; el León CUPRA, que aúna conducción ágil y divertida con eficiencia, hasta el León X-PERIENCE, con una versatilidad sin límites. Para en la familia del León el diseño ha sido uno de los factores clave en la decisión de compra, junto con la fiabilidad, la conducción ágil y divertida, el con-sumo y la conectividad. Las numerosas cualidades del León lo han convertido en un referente y en el modelo a seguir en cada producto con el logotipo SEAT.

DE Der Leon der aktuellen Generation ist ein außerordentlicher Erfolg und bewegt die gesamte Marke SEAT auf eine neue Stufe: Für Design und Qualität, Dynamik und Nutzwert, Technologie und Zuverlässigkeit gibt es Bestnoten von den Kunden – und von der Fachpresse: Bei rund 75 Prozent aller europaweiten Zeitschriften-Vergleichstests stand der Leon auf dem Podium.

Inzwischen ist die Leon-Familie zum zweiten starken Pfeiler der Marke geworden. In seiner Vielfalt spricht der Leon eine breite Zielgruppe an: vom Leon Ecomotive, der extreme Effizienz mit Fahrspaß verbindet, bis zum Leon CUPRA, der extremen Fahrspaß mit Effizienz verbindet, oder dem Leon X-PERIENCE, der in seiner Vielseitigkeit kaum Grenzen kennt. Design ist für die Zielgruppe der Leon-Familie immer ein entscheidender Kaufgrund, ebenso wie Zuverlässigkeit, Fahrspaß, Ökonomie oder Connectivity. Dabei ist der Leon mit seinen Qualitäten zum Maßstab und zum Vorbild für jedes Produkt mit dem SEAT Logo geworden.

FUTURE BRAND ICON SUV

THE BRAND PILLARS

IBIZA
LEON
SUV

EN SEAT goes SUV. The Spanish brand is continuing on its growth trajectory with the launch of its first SUV model in 2016, taking it into a new and important segment. According to market research, SUV sales in Europe are set to increase by a further third by 2018. The target group showing the strongest growth is that of slightly older customers, who particularly value the high seating position and an SUV's subjective feeling of safety. Equally important are middle-aged customers seeking a change in their own "auto-biography".

SEAT will clearly display its brand DNA in the SUV segment, too. Expressive design, great fun to drive and the very best quality will characterise SEAT's future SUV models just as much as their premium look-and-feel and excellent efficiency.

ES SEAT se pasa al SUV. La marca española continúa su trayectoria de crecimiento con el lanzamiento de su primer modelo SUV en 2016, y realiza así su debut en un segmento nuevo e importante. Según un estudio de mercado, para el año 2018 se prevé un crecimiento de más de un tercio en el mercado de los SUV en Europa. El grupo de población que muestra el mayor crecimiento es el de los clientes en la franja de los 40 y 50 años, que valoran particularmente la posición elevada del asiento y la sensación subjetiva de seguridad que proporciona el SUV. Igualmente importantes son los clientes de mediana edad que buscan un cambio en su «historia personal».

SEAT imprimirá también el ADN de la marca en el segmento de los SUV. Un diseño expresivo, una conducción divertida y la máxima calidad serán las características de los futuros modelos SUV de SEAT, junto con un aspecto exclusivo y una excelente eficiencia.

DE SEAT goes SUV: Mit der Einführung des ersten SUV-Modells im Jahr 2016 führt die spanische Marke ihren Wachstumskurs fort und besetzt ein wichtiges, neues Segment: Laut Marktforschung werden die SUV-Verkäufe in Europa bis zum Jahr 2018 um ein weiteres Drittel ansteigen. Die am stärksten wachsende Zielgruppe sind dabei die etwas älteren Kunden, sie schätzen die hohe Sitzposition und das subjektive Sicherheitsgefühl eines SUV besonders. Ebenso wichtig sind Kunden mittleren Alters, die in ihrer eigenen „Auto-Biografie" nach Abwechslung suchen.

Auch im SUV-Segment wird SEAT seine Marken-DNA klar zeigen: Expressives Design, großer Fahrspaß und beste Qualität werden die künftigen SUV-Modelle von SEAT ebenso auszeichnen wie hochwertige Anmutung und gute Effizienz.

SEAT
AN ATTRACTIVE EUROPEAN BASED BRAND

ES   SEAT – UNA MARCA EUROPEA ATRACTIVA
DE   SEAT – EINE ATTRAKTIVE MARKE MIT EUROPÄISCHEN WURZELN

EN   **SEAT's home is Spain and its heart belongs to southern Europe. Yet more than 80 percent of its production is exported. Its top markets include Germany, the United Kingdom, France and Italy. SEAT is represented in a total of 76 countries worldwide.**

ES   El hogar de SEAT está en España y su corazón pertenece al sur de Europa. Pero más del 80 por ciento de su producción se exporta. Sus mercados principales son Alemania, Reino Unido, Francia e Italia. SEAT está representada en un total de 76 países de todo el mundo.

DE   SEAT ist in Spanien zu Hause und hat sein Herz im Süden Europas. Dennoch gehen mehr als 80 Prozent der Produktion in den Export. Zu den Top-Märkten gehören Deutschland, Großbritannien, Frankreich und Italien. Insgesamt ist SEAT in 76 Ländern der Erde vertreten.

1,800
DEALERS
ES   CONCESIONARIOS
DE   HÄNDLERBETRIEBE

2,150
POINTS OF SALE
ES   PUNTOS DE VENTA
DE   VERKAUFSPUNKTE

2,900
SERVICE CENTRES
ES   CENTROS DE SERVICIO
DE   SERVICE-CENTER

GERMANY	BULGARIA	EL SALVADOR	ISRAEL
BELGIUM	ROMANIA	GUATEMALA	JORDAN
CZECH REPUBLIC	CROATIA	CUBA	MOROCCO
DENMARK	SLOVENIA	PANAMA	ALGERIA
FINLAND	MALTA	GUADELOUPE	TUNISIA
NORWAY	REPUBLIC OF SLOVAKIA	MARTINIQUE	EGYPT
SWEDEN	ALBANIA	RÉUNION	ANGOLA
FRANCE	MACEDONIA	PUERTO RICO	
UK	ESTONIA	ECUADOR	
IRELAND	LATVIA	PARAGUAY	
ICELAND	LITHUANIA	HONDURAS	
AUSTRIA	BOSNIA/HERZEGOVINA	TRINIDAD AND TOBAGO	
SWITZERLAND	GEORGIA	YEMEN	
ITALY	SERBIA	BAHRAIN	
SPAIN	MONTENEGRO	KUWAIT	
PORTUGAL	UKRAINE	OMAN	
POLAND	MEXICO	QATAR	
HUNGARY	CHILE	SAUDI ARABIA	
GREECE	DOMINICAN REPUBLIC	ABU DHABI	
TURKEY	COLOMBIA	DUBAI	
CYPRUS	VENEZUELA	UNITED ARAB EMIRATES	
NETHERLANDS	PERU	LEBANON	
LUXEMBOURG	BOLIVIA	AZERBAIJAN	

80

80 PERCENT OF PRODUCTION IS EXPORTED TO 76 COUNTRIES WORLDWIDE. THIS MAKES SEAT SPAIN'S MOST IMPORTANT INDUSTRIAL EXPORTER.

ES   EL 80 POR CIENTO DE LA PRODUCCIÓN SE EXPORTA A 76 PAÍSES DE TODO EL MUNDO. SEAT ES EL EXPORTADOR MÁS IMPORTANTE DEL SECTOR INDUSTRIAL ESPAÑOL.

DE   80 PROZENT EXPORTANTEIL, IN 76 LÄNDER WELTWEIT. DAMIT IST SEAT DER WICHTIGSTE INDUSTRIELLE EXPORTEUR SPANIENS.

SEAT
OUR
HOME
MARTORELL

THE FACTORY

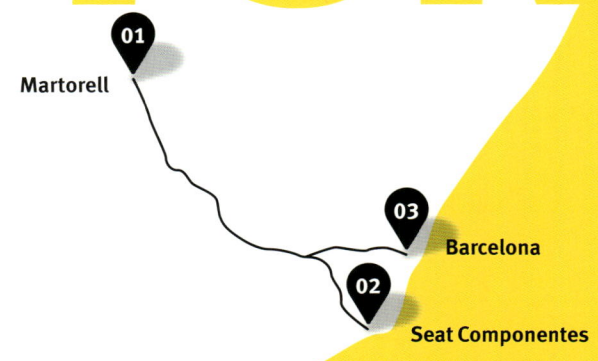

01   Martorell

03   Barcelona

02   Seat Componentes

2,100

VEHICLES PER DAY ACROSS FOUR PRODUCT RANGES

ES   VEHICULOS POR DIA DE CUATRO MODELOS DISTINTOS

DE   FAHRZEUGE PRO TAG AUS VIER PRODUKTFAMILIEN

EN   In Martorell, a new car is completed every 30 seconds – adding up to around 2,100 vehicles per day across four product ranges. Systematic training and development of each and every employee assures the outstanding quality typical of SEAT. There is only one way – the way to perfection.

ES   De Martorell sale un coche nuevo cada 30 segundos, lo que equivale a 2.100 vehículos al día entre las cuatro gamas de productos. La formación y la capacitación sistemática de cada empleado aseguran la alta calidad en SEAT. Solo hay un camino, el camino hacia la perfección.

DE   In Martorell wird alle 30 Sekunden ein neues Automobil fertiggestellt, rund 2.100 Fahrzeuge pro Tag aus vier Produktfamilien. Konsequente Weiterbildung und Training jedes Mitarbeiters sichern die für SEAT typische ausgezeichnete Qualität: Es gibt nur eine Richtung, den Weg zur Perfektion.

01 MARTORELL

EN   The plant in Martorell is one of the most modern production facilities in the global automotive industry – it was presented with the European Automotive Lean Award in 2013 for outstanding systems and processes.

ES   La fábrica de Martorell es una de las plantas de producción más modernas de la industria del automóvil. En 2013 fue galardonada con el prestigioso European Automotive Lean Award por la calidad y eficiencia de sus procesos productivos.

DE   Das Werk in Martorell ist eine der modernsten Fertigungsstätten der Automobilindustrie weltweit – 2013 wurde sie mit dem Lean Award für herausragende Prozesse und Abläufe ausgezeichnet.

02 SEAT COMPONENTES

EN   SEAT owns a modern transmission plant. It is located close to Barcelona Airport and produces transmissions for SEAT as well as other Volkswagen Group brands.

ES   SEAT cuenta con una moderna planta de transmisiones situada cerca del aeropuerto de Barcelona que fabrica transmisiones para SEAT y para otras marcas del Grupo Volkswagen.

DE   Zu SEAT gehört ein modernes Getriebewerk. Es liegt in der Nähe des Flughafens von Barcelona und produziert Getriebe für SEAT sowie für die anderen Marken des Volkswagen-Konzerns.

03 BARCELONA

EN   SEAT operates a modern component plant in Barcelona, supplying primarily pressed parts and subassemblies to the production facility in Martorell.

ES   SEAT cuenta con una moderna planta de componentes en Barcelona que suministra principalmente piezas prensadas y subensamblajes a la fábrica de producción de Martorell.

DE   In Barcelona betreibt SEAT ein modernes Komponentenwerk, das vor allem Pressteile und Vormontagen für die Produktion in Martorell zuliefert.

EN The Bodyshell Shop is home to highly flexible framer workstations and more than 2,000 robots. They carry out more than three billion spot welds per year with extreme precision. Laser measurement stations meticulously check quality down to tenths of a millimetre.

ES En el Taller de chapistería se encuentran estaciones de trabajo altamente flexibles y 2.000 robots que realizan más de 3.000 millones de soldaduras por puntos al año con la máxima precisión. Las estaciones de medición por láser comprueban la calidad de las operaciones con márgenes de décimas de milímetro.

DE Im Rohbau arbeiten hochflexible Framer-Workstations und mehr als 2.000 Roboter, sie setzen mit extremer Genauigkeit mehr als drei Milliarden Schweißpunkte im Jahr. Laser-Messstationen prüfen die Qualität auf den Zehntelmillimeter genau.

7,500

PEOPLE WORK AT THE MARTORELL PLANT. THEY ARE FULLY DEDICATED TO THE IMPROVEMENT OF QUALITY AND EFFICIENCY.

ES TRABAJADORES EN LA FÁBRICA DE MARTORELL ESTÁN DEDICADOS POR COMPLETO A LA MEJORA DE LA CALIDAD Y LA EFICIENCIA.

DE MITARBEITER ARBEITEN IN DER FABRIK MARTORELL STÄNDIG AN DER VERBESSERUNG VON QUALITÄT UND EFFIZIENZ.

5,000 m²

OF PRODUCTION SPACE ARE USED AS EMPLOYEE TRAINING AND DEVELOPMENT CENTRES. SEAT INVESTS AN AVERAGE OF 14 MILLION EUROS PER YEAR IN THE TRAINING AND DEVELOPMENT OF ITS PEOPLE FOR CONSISTENTLY HIGH PRODUCT QUALITY.

ES DE SUPERFICIE DE PRODUCCIÓN SE DESTINAN A LA FORMACIÓN DE EMPLEADOS Y A CENTROS DE DESARROLLO. SEAT INVIERTE UN PROMEDIO DE 14 MILLONES DE EUROS AL AÑO EN LA FORMACIÓN Y EL DESARROLLO DE SUS EMPLEADOS PARA GARANTIZAR LA MÁXIMA CALIDAD DE SUS PRODUCTOS.

DE WERDEN IN DER PRODUKTION ALS AUSBILDUNGS- UND TRAININGSZENTREN GENUTZT. SEAT INVESTIERT DURCHSCHNITTLICH 14 MILLIONEN EURO JÄHRLICH IN AUSBILDUNG UND TRAINING DER MITARBEITER FÜR KONSEQUENT HÖCHSTE PRODUKTQUALITÄT.

170

YOUNG EMPLOYEES, ON AVERAGE, ARE EDUCATED IN A DUAL VOCATIONAL TRAINING SYSTEM BASED ON THE GERMAN APPROACH.

ES APRENDICES RECIBEN FORMACIÓN PROFESIONAL DUAL BASADA EN EL MODELO ALEMÁN.

DE JUNGE MITARBEITER WERDEN DURCHSCHNITTLICH IN EINEM DUALEN SYSTEM NACH DEUTSCHEM VORBILD AUSGEBILDET.

SEAT AL SOL

SUSTAINABILITY

It is not only SEAT cars that lead the way in efficiency; the factory in Martorell owns the largest solar power facility in the global automotive industry. Around 53,000 solar panels on an area of 276,000 m² capture the Spanish sunlight and supply the production plant in Martorell with energy. In 2014, 17 million kWh of environmentally friendly electricity – enough to meet the needs of 4,000 families. Some of the panels also function as roofing for parking areas, providing protection for the new cars.

SOSTENIBILIDAD

No solo son los coches SEAT los que marcan el camino en cuanto a ahorro energético. La fábrica de Martorell gestiona la mayor instalación fotovoltaica de la industria del automóvil. Cuenta con casi 53.000 paneles solares que ocupan una superficie de 276.000 m² y abastecen de energía limpia a la planta de producción de Martorell. En 2014 los paneles generaron 17 millones de kWh de electricidad limpia, suficientes para satisfacer la demanda de 4.000 hogares. Algunos de los paneles cumplen la doble función de proteger los coches de las inclemencias del tiempo ya que están instalados en las campas donde se encuentran los vehículos ya fabricados.

NACHHALTIGKEIT

Die Sonne ist unser Partner. Nicht nur die Automobile von SEAT sind führend in Sachen Effizienz: Die Fabrik in Martorell besitzt die in der Automobilindustrie weltweit größte Solaranlage, SEAT al Sol: Auf 276.000 m² Fläche fangen rund 53.000 Solarpaneele das spanische Sonnenlicht ein und versorgen das Fertigungswerk in Martorell mit Energie. 2014 erzeugten die Paneele 17 Millionen kWh umweltfreundlichen Strom – genug um den Energiebedarf von 4.000 Haushalten zu decken. Zudem fungiert ein Teil der Paneele als Parkplatzüberdachung und schützt die Neuwagen.

17

MILLION KWH OF SOLAR ENERGY PER YEAR

<u>ES</u> MILLONES DE KWH DE ENERGÍA SOLAR AL AÑO

<u>DE</u> MILLIONEN KWH AUS SOLARENERGIE PRO JAHR

SEAT
TECHNICAL CENTER RESEARCH AND DEVELOPMENT

EN Around 1,000 engineers work at SEAT's Technical Center in Martorell on the future of mobility, securing the brand's position at the forefront of technology. On a facility covering around 43,000 m², new models are conceived using state-of-the-art systems and equipment. It also undertakes projects for the Volkswagen Group. Around 100 people work at SEAT's Design Center, shaping cars with emotional styling and the very highest precision. They have developed a distinctive design language that is recognised and acknowledged all over the world.

ES En el Centro Técnico de SEAT en Martorell trabajan alrededor de 1.000 ingenieros en el futuro de la movilidad que garantizan el posicionamiento de la marca a la vanguardia tecnológica. Los nuevos modelos se conciben en unas instalaciones decasi 43.000 m² dotadas de los sistemas y equipamiento más modernos. También realizan proyectos para el Grupo Volkswagen. Alrededor de 100 personas trabajan en el Centro un diseño de SEAT en el desarrollo de coches con impacto emocional y la mayor precisión. SEAT ha creado un lenguaje de diseño inconfundible que se reconoce y se aprecia en todo el mundo.

DE Im Centro Tecnico von SEAT in Martorell arbeiten rund 1.000 Ingenieure an der Zukunft der Mobilität und sichern die technologische Spitzenposition der Marke. Auf rund 43.000 m² Gebäudefläche werden die neuen Modelle mit modernsten Anlagen konzipiert und zusätzlich Projekte für den gesamten Volkswagen-Konzern bearbeitet. Rund 100 Mitarbeiter gestalten im Design Center Automobile mit emotionalen Formen und höchster Präzision – sie haben eine unverwechselbare und weltweit hoch anerkannte Formensprache entwickelt.

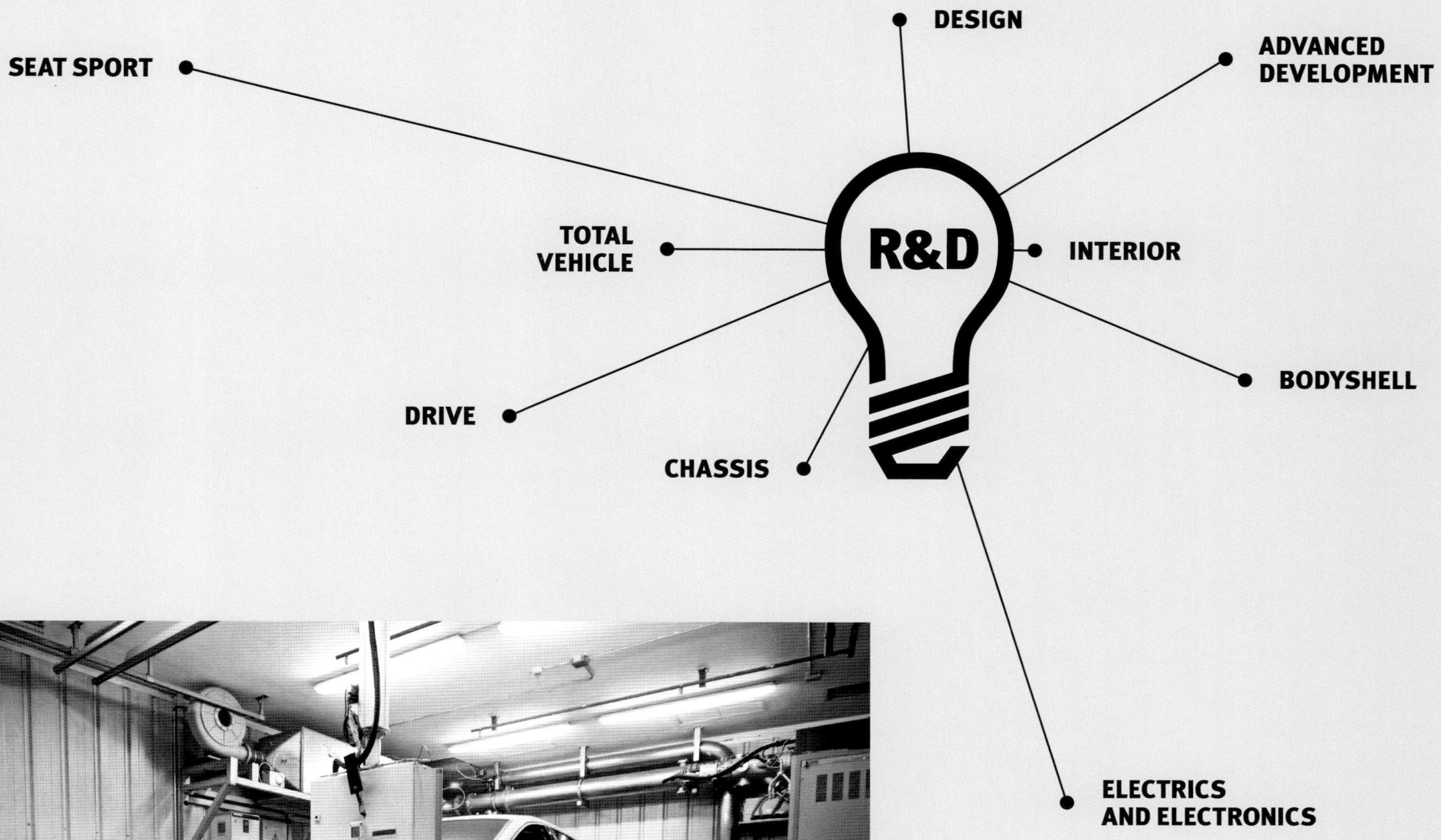

SEAT SPORT •

• DESIGN

• ADVANCED
DEVELOPMENT

R&D

TOTAL
VEHICLE •

• INTERIOR

DRIVE •

• BODYSHELL

CHASSIS •

ELECTRICS
AND ELECTRONICS

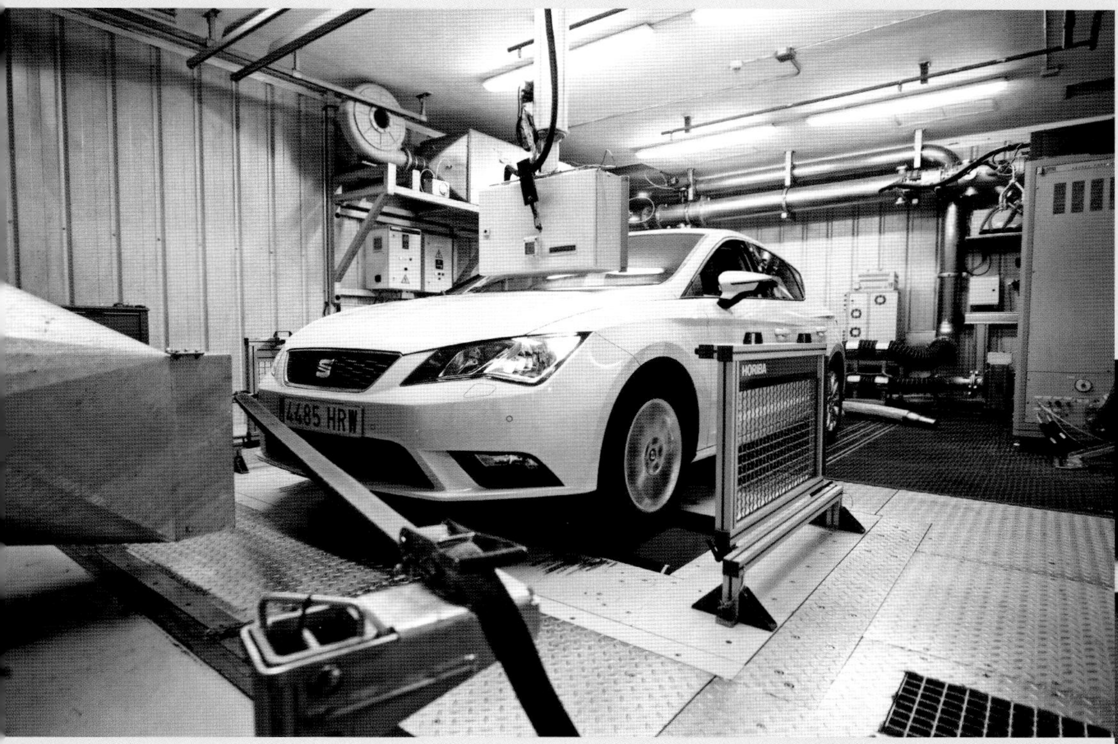

1,000

AROUND 1,000 ENGINEERS WORK AT SEAT'S TECHNICAL CENTER.

<u>ES</u> ALREDEDOR DE 1.000 INGENIEROS TRABAJAN EN
EL CENTRO TÉCNICO DE SEAT.

<u>DE</u> RUND 1.000 INGENIEURE ARBEITEN IM CENTRO TECNICO
VON SEAT.

SEAT THE TEAM

WE PRODUCE DRIVING FUN AND PRIDE OF OWNERSHIP

EN It's the people who make the company – the employees who bring the brand to life. More than 14,000 men and women say "We are SEAT". They are proud to work with all their skill and their passion on great products.

ES La empresa la hacen las personas. Son los empleados quienes dan vida a la marca. Más de 14.000 hombres y mujeres afirman «Nosotros somos SEAT». Están orgullosos de destinar toda su pasión y hacer uso de sus capacidades para crear buenos productos.

DE Die Menschen machen das Unternehmen, die Mitarbeiter bringen die Marke zum Leben. „Wir sind SEAT", sagen mehr als 14.000 Frauen und Männer. Sie sind stolz darauf, mit all ihrem Können und ihrer Leidenschaft an tollen Produkten zu arbeiten.

MANAGING DIRECTOR OF SEAT GERMANY
AND AT SEAT SINCE
2014

SEAT is very successful, especially in Germany. A growing brand reputation, excellent products and very active dealers ensure that sales figures continue to grow in one of the most demanding and hard-fought car markets in the world. With the expansion of the dealer network in terms of the number and quality of the facilities, Bernhard Bauer is establishing one of the most important bases for ongoing success. As the Head of SEAT Germany, he knows that customers expect flexible service tailored to their individual needs. "Another key to the German market is in the broad model line-up with its diverse opportunities for individualisation."

Bernhard Bauer has worked for SEAT for just one year. But he was impressed from day one by the innovative thinking of his colleagues and by their target of continually overcoming boundaries. "Being part of this family and making SEAT a little better every single day – that's what I'm fighting for".

ES CARGO
DIRECTOR GENERAL DE SEAT ALEMANIA
EN SEAT DESDE
2014

SEAT es todo un éxito, especialmente en Alemania. La mejora de reputación de la marca, coches sobresalientes y concesionarios muy activos garantizan que las cifras de ventas continúen creciendo en uno de los mercados más exigentes y competitivos del mundo. Gracias a la ampliación de la red de concesionarios tanto en número como en calidad de las instalaciones, Bernhard Bauer está estableciendo bases sólidas para un éxito continuado. Como director de SEAT Alemania, sabe que los clientes esperan un servicio flexible y adaptado a sus necesidades individuales. «Otra de las características clave del mercado alemán es la amplia oferta de modelos y las grandes posibilidades de personalización».

Bernhard Bauer apenas lleva un año trabajando en SEAT, pero ya desde el primer día le impresionó la mentalidad innovadora de sus compañeros y el empeño para superar las dificultades. «Formar parte de esta familia y hacer que SEAT sea cada día mejor es mi principal objetivo».

DE POSITION
GESCHÄFTSFÜHRER SEAT DEUTSCHLAND
BEI SEAT SEIT
2014

SEAT ist sehr erfolgreich, gerade auch in Deutschland. Steigendes Ansehen der Marke, exzellente Produkte und sehr aktive Händler sorgen für beständig wachsende Verkaufszahlen auf einem der anspruchsvollsten und am härtesten umkämpften Automobilmärkte der Welt. Mit dem Ausbau des Vertriebsnetzes in Zahl und Qualität der Betriebe schafft Bernhard Bauer eine der wichtigsten Grundlagen für den weiteren Erfolg. Als Chef von SEAT Deutschland weiß er, dass die Kunden einen flexiblen, zu ihren individuellen Ansprüchen passenden Service erwarten. „Ein weiterer Schlüssel für den deutschen Markt liegt in der breiten Modellpalette mit ihren vielfältigen Möglichkeiten zur Individualisierung."

Bernhard Bauer arbeitet erst seit einem Jahr für SEAT. Aber er ist seit dem ersten Tag begeistert vom innovativen Denken der Kollegen und von ihrem Ziel, stets Grenzen zu überwinden. „Teil dieser Familie zu sein und SEAT jeden Tag etwas besser zu machen – dafür kämpfe ich."

REGARDLESS OF THEIR AGE, PEOPLE WHO DRIVE A SEAT ARE MORE INDEPENDENT, MORE INQUISITIVE AND MORE UNCONVENTIONAL THAN OTHERS.

ES Sin tener en cuenta su edad, las personas que conducen un SEAT son más independientes, curiosas y menos convencionales que otras.
DE In jedem Alter sind Menschen, die einen SEAT fahren, unabhängiger, neugieriger und unkonventioneller als andere.

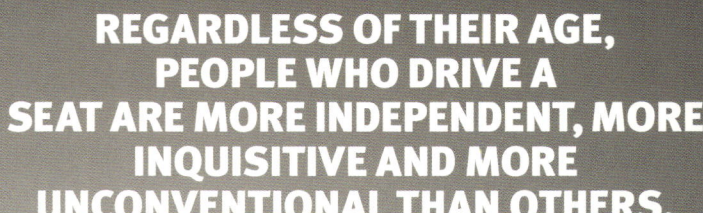

MANAGING DIRECTOR OF SEAT GERMANY

BERNHARD BAUER

ANDRÉS ASENSIO

WE GUARANTEE THAT EVERY CAR EXCEEDS CUSTOMER EXPECTATIONS, FROM THE VERY FIRST ONE SOLD.

ES Garantizamos que todos los vehículos superen las expectativas de los clientes, desde el primero vendido.
DE Wir garantieren, dass schon das erste verkaufte Auto die Erwartungen des Kunden übertrifft.

EN POSITION
PROTOTYPE MANUFACTURING
AT SEAT SINCE
1999

This is where ideas take shape; this is where we turn plans into reality. Andrés Asensio and his colleagues take sheet metal and transform it into the very first examples of a car never seen before: They are the people who build the diverse array of prototypes for new models. These prototypes are used for further development, for rigorous testing, for production planning and for assuring maximum quality.

This is where something new is created and, even if it takes months before the first unit is handed over to its new owner, the customer is always the focal point. For every one of these extremely complex and sophisticated prototypes, one question remains uppermost: What can we do better? How can we raise the perceived quality and the reliability yet another notch higher? At the end of the day, the very first car delivered to its customer must be nothing less than perfect.

ES CARGO
CREACIÓN DE PROTOTIPOS
EN SEAT DESDE
1999

Aquí es donde las ideas toman forma. Es donde convertimos los planes en realidad. Andrés Asensio y sus compañeros toman unas hojas de metal y las transforman en los primeros ejemplares de un vehículo inexistente hasta el momento. Personas que crean un abanico de prototipos para nuevos modelos. Estos prototipos se usan para poner a punto el proyecto con las pruebas más exigentes, a fin de asegurar la planificación de la producción y garantizar la máxima calidad.

Aquí es donde se crea algo nuevo y, aunque tengan que pasar meses antes de que la primera unidad llegue a manos de su nuevo propietario, el cliente siempre está en el objetivo. Con cada uno de los prototipos, extremadamente complejos y sofisticados, hay una cuestión que nos preocupa principalmente: ¿En qué podemos mejorar? ¿Cómo podemos aumentar todavía más la calidad percibida y la fiabilidad? En última instancia, lo importante es que el primer coche que se entregue al cliente sea como mínimo perfecto.

DE POSITION
PROTOTYPEN-FERTIGUNG
BEI SEAT SEIT
1999

Hier nehmen die Ideen Form an, hier wird aus Plänen Realität. Andrés Asensio verwandelt mit seinen Kollegen schlichte Blechtafeln in die ersten Exemplare eines noch nie gesehenen Automobils: Sie bauen die vielfältigen Prototypen der neuen Modelle. Diese werden gebraucht für die weitere Entwicklung, für harte Tests, für die Produktionsplanung und für die Sicherung maximaler Qualität.

Hier wird Neues geschaffen, und selbst wenn es noch Monate dauert, bis das erste Exemplar seinem Käufer übergeben wird: Der Kunde steht dabei im Mittelpunkt. Denn bei jedem einzelnen der extrem aufwendigen Prototypen geht es um die Frage: Was können wir besser machen? Wie können wir die Anmutung oder die Zuverlässigkeit noch steigern? Schließlich muss schon das erste Kundenauto nichts weniger sein als perfekt.

POSITION
NETWORK DEVELOPMENT
AT SEAT SINCE
2003

The salesperson is the face of the brand to our customers. Ensuring a positive experience with our brand and being passionate about high working standards are key aspects for cultivating customer loyalty. These include everything from the initial contact with customers when making a sale to the different after-sales services we provide.

Ferran Jover works in the Sales Network Development Department, and part of his responsibilities lie in selecting the best commercial partners for SEAT. "We have to constantly enhance our contact with customers and ensure the best representation in the markets, two keys points that significantly contribute to SEAT's sustainable growth".

Ferran joined SEAT in the Trainee programme in 2003, and he is just as aware today of the professional challenges his job offers as he was back then. "SEAT offers many possibilities. Working with a group of people who constantly add value to the company makes you strive to do your best every day".

ES CARGO
DESARROLLO RED
EN SEAT DESDE
2003

El comercial es la cara visible de la marca frente a nuestros clientes. El asegurar una buena experiencia con nuestra marca y la pasión por la calidad en el trabajo son las claves para fidelizar a nuestros clientes. Abarcan desde el primer contacto para la venta, hasta los diversos servicios que se prestan en la postventa.

Ferran Jover trabaja en el Departamento de Desarrollo de Red Comercial, y entre su actividades está la selección de los mejores socios comerciales para SEAT. "Debemos mejorar de forma continua el contacto con los clientes y asegurar la mejor representación en los mercados, dos puntos clave que contribuyen de forma importante al crecimiento sostenible de SEAT."

Ferran comenzó en SEAT en 2003 en el programa Trainee, y hoy valora como el primer día los retos profesionales que le ofrece su trabajo. "SEAT ofrece muchísimas posibilidades. Trabajar rodeado de personas que constantemente añaden valor a la empresa hace que intentes dar lo mejor de ti cada día".

DE POSITION
HÄNDLERNETZ-ENTWICKLUNG
BEI SEAT SEIT
2003

Der Verkäufer ist das Gesicht der Marke gegenüber dem Kunden. Eine positive Erfahrung zu sichern und mit Passion an einem hohen Standard zu arbeiten, ist entscheidend für die Loyalität der Kunden. Das ist wichtig im gesamten Prozess, vom ersten Kontakt zum Kunden über den Verkauf bis zu den verschiedenen After-Sales-Services, die SEAT anbietet.

Ferran Jover arbeitet in der Händlernetz-Entwicklung. Zu seinen Aufgaben zählt es, für SEAT die besten Partner auszuwählen. "Wir müssen unseren Kontakt zu den Kunden beständig verbessern und die bestmögliche Vertretung im Markt sicherstellen. Für das nachhaltige Wachstum von SEAT sind das zwei entscheidende Punkte."

Ferran kam 2003 zu SEAT und begann in einem Trainee-Programm. Aber noch heute empfindet er seine beruflichen Herausforderungen ebenso spannend wie damals. "SEAT bietet viele Möglichkeiten. Wenn Du mit Menschen zusammen bist, die den Wert der Marke jeden Tag steigern, dann spornt Dich das auch zu Höchstleistungen an."

SEAT IS A COMPANY THAT OFFERS FASCINATING CHALLENGES AND ENABLES YOU TO SHARE THEM WITH GREAT PROFESSIONALS ON A DAILY BASIS.

ES SEAT es una empresa que ofrece retos fascinantes y permite compartirlos con grandes profesionales cada día.
DE SEAT ist ein Unternehmen, das faszinierende Herausforderungen bietet. Bearbeiten kann man sie gemeinsam mit höchst professionellen Kollegen.

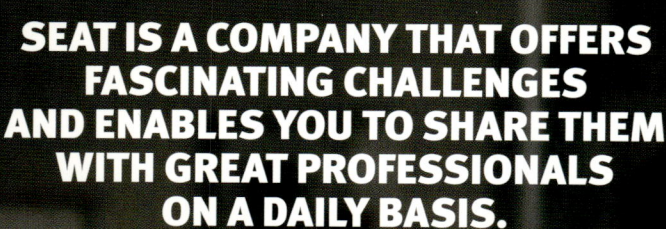

NETWORK DEVELOPMENT

FERRAN JOVER

POSITION
SALES EUROPE
AT SEAT SINCE
1986

María Luisa Santa Coloma loves to travel, meet new people and experience new cultures. And she has the good fortune that her job at SEAT fits in perfectly with this passion. Within the Sales department in Martorell, she takes care of the European markets, where the brand has been highly successful for some considerable time, with admirable growth rates in countries such as Germany, Italy, the Czech Republic and Turkey. This positive momentum is something that must be maintained and expanded.

"I'm very happy to be able to work with different countries. Even though my direct 'customers' are the national sales organisations, we are constantly thinking about all the potential customers, buyers and drivers of a SEAT," says María Luisa. "At the end of the day, we have to make sure that their wishes and ideas are completely fulfilled." And these are often incredibly varied, even within Europe.

ES **CARGO**
VENTAS EUROPA
EN SEAT DESDE
1986

A María Luisa Santa Coloma le apasiona viajar, conocer gente y nuevas culturas. Tiene la suerte de que su trabajo en SEAT encaja perfectamente con esa pasión. María Luisa es responsable de varios mercados europeos en el Departamento de Ventas de Martorell. En los últimos meses la marca está teniendo mucho éxito con cifras de crecimiento destacadas en mercados como Alemania, Italia, República Checa y Turquía. Un éxito que debe mantenerse y desarrollarse.

«Tengo la suerte de poder trabajar con varios países. Aunque mis «clientes» directos son las organizaciones nacionales, todos tenemos siempre en mente los intereses del comprador y conductor de un vehículo SEAT», dice María Luisa. «En definitiva, intentamos que se cumplan todos sus deseos y expectativas». Y ciertamente son muy distintos entre países, incluso dentro de Europa.

DE **POSITION**
VERTRIEB EUROPA
BEI SEAT SEIT
1986

María Luisa Santa Coloma liebt es zu reisen und dabei Menschen und Kulturen kennenzulernen. Und sie hat das Glück, dass ihre Aufgabe bei SEAT perfekt zu dieser Leidenschaft passt: Im Vertrieb in Martorell betreut sie die europäischen Märkte. Und gerade da ist die Marke seit geraumer Zeit sehr erfolgreich, mit beachtlichen Wachstumsraten beispielsweise in Deutschland, Italien, der Tschechischen Republik oder auch der Türkei. Ein Erfolgsmomentum, das es zu bewahren und auszubauen gilt.

„Ich bin glücklich, mit verschiedenen Ländern zusammenarbeiten zu können. Auch wenn meine direkten ‚Kunden' die nationalen Vertriebsorganisationen sind, haben wir immer und bei allem die Interessenten, Käufer und Fahrer eines SEAT im Sinn", sagt María Luisa. „Schließlich müssen wir dafür sorgen, dass deren Wünsche und Vorstellungen exakt erfüllt werden." Und die sind oft recht unterschiedlich, auch innerhalb Europas.

WE DO EVERYTHING WE CAN TO MAKE PEOPLE PROUD TO DRIVE A SEAT. A SATISFIED CUSTOMER IS THE BEST AMBASSADOR THE BRAND CAN HAVE.

ES Hacemos todo lo posible para que nuestros clientes se sientan orgullosos de conducir un SEAT. Un cliente satisfecho es el mejor embajador de la marca.
DE Wir tun alles dafür, dass die Menschen stolz sind, einen SEAT zu fahren. Ein zufriedener Kunde ist der beste Botschafter der Marke.

SALES EUROPE

MARÍA LUISA SANTA COLOMA

Leyre Olavarría is constantly connected – with the digital world, with the present and with the future. She works in the Centro Tecnico in Martorell on one of the most decisive technology issues of the next few years – the networking of the automobile and its driver with the outside world. Connectivity is taken for granted as part of a young, urban lifestyle and SEAT holds a leading position. FullLink creates the perfect smartphone connection, while the SEAT ConnectApp offers exclusive features.

"We want to maintain this lead. It's very important to our customers," says Leyre Olavarría. It's a lead based on the close cooperation with Samsung, but above all, on the creativity of the in-house team. "We are incredibly dynamic at SEAT. And we have the drive to be right at the front of the pack."

Leyre Olavarría ist immer in Verbindung – mit der digitalen Welt, mit der Gegenwart, mit der Zukunft. Sie arbeitet im Centro Tecnico in Martorell an einem der entscheidenden Technologiethemen der nächsten Jahre: der Vernetzung des Automobils und seines Fahrers mit der Umwelt. Connectivity gehört ganz selbstverständlich zum jungen, urbanen Lifestyle, und SEAT hat da eine führende Position. FullLink schafft die perfekte Verbindung zum Smartphone, die SEAT ConnectApp bietet exklusive Features.

„Wir wollen diesen Vorsprung behalten, unsere Kunden schätzen das sehr", weiß Leyre Olavarría und baut auf die enge Kooperation mit Samsung, vor allem aber auf die Kreativität des eigenen Teams: „SEAT ist ein sehr dynamisches Unternehmen. Und wir haben den Willen, ganz vorne zu sein."

Leyre Olavarría está constantemente conectada; con el mundo digital, con el presente y con el futuro. Trabaja en el Centro Técnico de Martorell, en una de las cuestiones tecnológicas más decisivas de los próximos años: la creación de redes que conecten el automóvil y a su conductor con el mundo exterior. La conectividad se considera parte de un estilo de vida joven y urbano, y SEAT es líder indiscutible en este campo. FullLink crea la conexión perfecta para dispositivos inteligentes, mientras que SEAT ConnectApp ofrece unas características exclusivas.

«Queremos mantener este liderazgo. Es muy importante para nuestros clientes», declara Leyre Olavarría. Es una posición destacada basada en la cooperación con Samsung, pero sobre todo, en la creatividad de nuestra plantilla. «En SEAT somos increíblemente dinámicos, y tenemos la energía y fuerza necesarias para estar a la cabeza».

I'M REALLY LUCKY TO BE WORKING IN AN EXTREMELY DYNAMIC COMPANY. EVERY PROJECT SETS US NEW CHALLENGES.

ES Soy muy afortunada por poder trabajar en una compañía tan extremadamente dinámica. Cada proyecto plantea nuevos retos.
DE Ich habe das große Glück, in einem sehr dynamischen Unternehmen zu arbeiten. Jedes Projekt stellt uns vor neue Herausforderungen.

CONNECTIVITY DEVELOPMENT

LEYRE OLAVARRÍA

PEDRO VALLEJO

THE CUSTOMER KNOWS NOTHING ABOUT TENTHS OF A MILLIMETRE. BUT THEY FEEL THE QUALITY!

ES El cliente no entiende de décimas de milímetro. ¡Pero percibe nuestra calidad!

DE Der Kunde weiß nichts von Zehntelmillimetern. Aber er fühlt die Qualität!

EN POSITION
MEISTERBOCK
AT SEAT SINCE
2009

This workstation is so German that there's no real translation for it. "Meisterbock" is the name given to the massive platform made from stainless steel and aluminium profiles, because this is where truly "masterful" work is carried out to the very highest standards.

The Meisterbock is the master on which every new SEAT model is based. And just as the Mètre des Archives defines our metric measurement system, the Meisterbock stands for the absolute precision of every single component. The people working here agonize over tenths of a millimetre, constantly in search of perfection. "When the customer sits in a SEAT, he senses the perfect interaction that we work so hard to achieve," says Pedro Vallejo. "They feel the quality. And that's what sets our work apart."

ES CARGO
MEISTERBOCK
EN SEAT DESDE
2009

Este taller es tan alemán que no es posible traducir su nombre. «Meisterbock» es la denominación con la que se ha bautizado a la estructura hecha de perfiles de aluminio y de acero inoxidable, porque aquí es donde se lleva a cabo el verdadero trabajo «maestro» conforme a los estándares más exigentes.

Meisterbock es el modelo patrón en el que se montan, miden y analizan todas las piezas de los nuevos modelos SEAT. Y al igual que el Mètre des Archives define nuestro sistema métrico de medición, el Meisterbock simboliza la precisión absoluta. Las personas que trabajan aquí analizan con gran esfuerzo y pasión cada décima de milímetro, ya que es el único camino para lograr la perfección. «Cuando el cliente entra en un SEAT, siente la perfecta interacción entre cada una de las piezas por la que tan intensamente trabajamos», dice Pedro Vallejo. «Se siente la calidad, y esto es lo que marca la diferencia de nuestro trabajo».

DE POSITION
MEISTERBOCK
BEI SEAT SEIT
2009

Dieser Arbeitsplatz ist so deutsch, dass es gar keine Übersetzung gibt: „Meisterbock" heißt das massive Podest aus Edelstahl und Aluminiumprofilen, denn hier wird wirklich „meisterlich" gearbeitet – auf allerhöchstem Niveau.

Der Meisterbock ist die Urform jedes neuen Modells von SEAT. Und so wie das Urmeter unser metrisches Maßsystem definiert, so steht der Meisterbock für die absolute Präzision jedes einzelnen Bauteils. Um Zehntelmillimeter wird hier gerungen, stets auf der Suche nach Perfektion. „Wenn der Kunde in einem SEAT sitzt, dann spürt er das perfekte Zusammenspiel", sagt Pedro Vallejo. „Er fühlt die Qualität. Und genau das ist das Ziel unserer Arbeit."

PATRICIA COLLADOS

YOU NOT ONLY HAVE TO FIND TALENTED PEOPLE, YOU ALSO HAVE TO WIN THEIR LOYALTY AND KEEP THEM MOTIVATED.

ES El objetivo no es solo encontrar personas con talento, sino ganarse su fidelidad y mantenerlas motivadas.

DE Talente muss man nicht nur finden. Man muss sie auch ans Unternehmen binden und immer wieder motivieren.

ES   CARGO
RECURSOS HUMANOS
EN SEAT DESDE
1997

Patricia Collados habla con mucha gente. Porque busca lo mejor de lo mejor. Y a los mejores, les ofrece los mejores puestos de la empresa. «Solo cuando nuestros empleados están orgullosos de su trabajo y orgullosos de SEAT, son capaces de contribuir a un auténtico éxito sostenible y a producir productos de máxima calidad».

Patricia está orgullosa de su compañía, «Porque nos ofrece los mejores recursos, como movilidad internacional, diversidad intercultural y grandes oportunidades de desarrollo personal». El amplio programa formativo de la empresa, que abarca desde estudios duales (que combinan trabajo y educación universitaria) hasta una vasta variedad de oportunidades de desarrollo profesional, sitúa a SEAT en una posición de liderazgo dentro de la economía española y de la industria automovilística en su conjunto.

EN   POSITION
HUMAN RESSOURCES
AT SEAT SINCE
1997

Patricia Collados talks to a lot of people – because she is seeking the very best. And for them, the very best jobs within the company. "Only when our employees are proud of their work and proud of SEAT, are they able to contribute to truly good products and sustainable success."

Patricia is herself proud of her company, "Because it gives us the best resources, like international mobility, a wealth of intercultural diversity and enormous opportunities for further personal development." The company's extensive training programme, which ranges from dual studies (combining work and university education) to wide-ranging career development opportunities, puts SEAT in a leading position within the Spanish economy and the automotive industry as a whole.

DE   POSITION
PERSONALMANAGEMENT
BEI SEAT SEIT
1997

Patricia Collados führt viele Gespräche. Denn sie sucht die Besten. Und für sie die besten Plätze im Unternehmen. „Nur wenn unsere Mitarbeiter stolz auf ihre Arbeit und stolz auf SEAT sind, tragen sie zu wirklich guten Produkten und zu nachhaltigem Erfolg bei."

Patricia selbst ist stolz auf ihr Unternehmen, „denn SEAT gibt uns optimale Ressourcen wie internationale Mobilität, interkulturellen Reichtum und intensive Möglichkeiten zur persönlichen Weiterbildung." Gerade mit dem umfangreichen Trainingsprogramm – von der Dualen Ausbildung bis hin zu vielfältigen Aufstiegsmöglichkeiten – nimmt SEAT eine führende Position in der spanischen Wirtschaft und auch in der Automobilindustrie ein.

Raul is a very particular kind of perfectionist. The paintwork must not merely shine, but veritably sparkle like a mirror under the Spanish sun. Every millimetre out of place is his personal enemy. He checks the exact fit of all parts with extreme precision. When it comes to bodyshell shutlines, he's looking for tenths of a millimetre.

At the end of the day, his vehicles stand quite literally in the spotlight. Raul prepares the cars for international motor shows, fairs and special events. He therefore presents the best possible business cards for the Spanish brand. His job is pure passion for him, because the quality of every single SEAT automobile is so good today that he can concentrate on searching for total perfection.

Raul es un perfeccionista muy particular. La pintura no debe simplemente brillar, tiene que resplandecer como un espejo bajo el sol. Cualquier décima de desviación en los ajustes se convierte en su enemigo personal. Comprueba que todas las piezas ajusten exactamente entre ellas hasta la última décima. Cuando se trata de franquicias de carrocería, las examina milímetro a milímetro.

Al final del día, sus vehículos están literalmente en el punto de mira. Raúl prepara los coches para salones internacionales del automóvil, exposiciones y eventos especiales. Por ello, saca a relucir las mejores tarjetas de presentación de la marca española. Tiene un trabajo que le apasiona porque, hoy en día, la calidad de todos los automóviles de SEAT es tan alta que se puede concentrar en buscar la perfección total.

Raul Guitián ist ein Perfektionist der ganz besonderen Art. Der Lack muss nicht nur glänzen, er muss leuchten wie ein Spiegel unter der spanischen Sonne. Jeder Millimeter Abweichung ist sein persönlicher Feind. Er prüft die Passung jedes Teils mit höchster Präzision, bei den Karosseriefugen geht es um Zehntel.

Schließlich stehen seine Fahrzeuge im sprichwörtlichen Sinne im Rampenlicht: Raul Guitián bereitet die Autos für die internationalen Automobilmessen, für Ausstellungen und besondere Events vor – als bestmögliche Visitenkarten für die spanische Marke. Für Raul ist sein Job pure Leidenschaft – und Freude: Denn die Qualität jedes einzelnen SEAT Automobils ist heute so perfekt, dass er sich auf den finalen Hauch von Perfektion konzentrieren kann.

WE WANT TO SHOW THE PERFECTION OF OUR PRODUCTS TO THE CUSTOMER. AND THAT'S EXACTLY WHAT WE DO.

ES Queremos mostrar al cliente la perfección de nuestros productos. Y eso es exactamente lo que hacemos.

DE Wir wollen den Kunden von der Perfektion unserer Produkte überzeugen. Und das gelingt uns auch.

SPECIAL CAR PREPARATION

RAUL GUITIÁN

EXTERIOR QUALITY

ANA LEMOS

EN POSITION
EXTERIOR QUALITY
<u>**AT SEAT SINCE**</u>
2012

Ana Lemos has a very sharp eye, and high-precision measuring equipment that supports her. As far as she is concerned, there are no tolerances. Lemos is responsible for the quality of exterior parts during the launch phase of new projects. She inspects the parts that come to the Martorell factory from several hundred suppliers.

"The typical SEAT design can only be achieved with absolute precision," she says. "Even a bumper has to fit to one tenth of a millimetre." It is for this reason that she is also involved in the selection of the best suppliers, who are then supported in a process of continual improvement in the development of the parts and their subsequent production. "Our attention to detail is what makes the difference," says Ana Lemos.

ES CARGO
CALIDAD EXTERIOR
<u>**EN SEAT DESDE**</u>
2012

Ana Lemos tiene un ojo de lince y cuenta con el apoyo de un equipo de medición de alta precisión. En lo que a ella respecta, no hay que tolerar la más mínima divergencia. Lemos es la encargada de la calidad de piezas de exterior en la fase de lanzamiento de nuevos proyectos. Inspecciona las piezas que llegan a la fábrica de Martorell procedentes de varios cientos de proveedores.

«El diseño típico de SEAT solo se puede lograr con la precisión absoluta», dice. «Incluso un parachoques debe estar ajustado hasta la más mínima décima de milímetro». Por ello, también está involucrada en la selección de los mejores proveedores, y en la posterior participación en un proceso de mejora continua en cuanto al desarrollo de piezas y su subsecuente producción. «Nuestra atención por los detalles es lo que marca la diferencia», declara Ana Lemos.

DE POSITION
EXTERIEUR QUALITÄT
<u>**BEI SEAT SEIT**</u>
2012

Ana Lemos hat ein scharfes Auge. Und hochpräzise Messinstrumente, die sie unterstützen. Bei ihr gibt es keine Toleranzen. Sie ist verantwortlich für die Qualität des Exterieurs in der Launchphase eines neuen Modells. Sie überprüft die Teile, die von mehreren hundert Zulieferern ins Werk nach Martorell kommen.

„Das typische Design von SEAT ist nur mit absoluter Präzision realisierbar", weiß Ana Lemos. „Selbst ein Stoßfänger muss auf den Zehntelmillimeter genau passen." Deshalb begleitet sie bereits die Auswahl der besten Lieferanten. Sie werden dann beim Prozess der ständigen Verbesserung der Teile in Entwicklung und Produktion unterstützt. Ana Lemos: „Es ist unsere Liebe zum Detail, die den Unterschied macht".

ONLY IF WE GIVE OUR VERY BEST EVERY DAY WILL WE CONSISTENTLY BE ABLE TO SET OURSELVES APART FROM THE COMPETITION.

ES Solo lograremos diferenciarnos de manera consistente de la competencia si cada día damos lo mejor de nosotros mismos.
DE Nur wenn wir jeden Tag unser Bestes geben, werden wir uns dauerhaft vom Wettbewerb unterscheiden.

ÓSCAR FERNÁNDEZ

ALL AREAS OF THE COMPANY ARE NOW FOCUSED ON QUALITY – AND THEREFORE ON THE CUSTOMER.

ES En la actualidad, todas las áreas de la empresa están centradas en la calidad, y por tanto, en el cliente.
DE Inzwischen konzentrieren sich alle Bereiche des Unternehmens auf die Qualität – und damit auf den Kunden.

EN **POSITION**
MEISTERBOCK
AT SEAT SINCE
2001

And back again to the "Meisterbock". This measurement device – which appears so simple, yet is, in detail, highly intelligent and extremely precise – is something of a development and production nerve centre in Martorell. The technicians responsible, like Óscar Fernández, call it a "zero-reference model". This is where every single part from the SEAT Press Shop and the Bodyshell Manufacturing in Martorell and Barcelona, and every single plastic part from the suppliers have to fit precisely to within hundredths of a millimetre.

The Meisterbock demonstrates how well SEAT combines the best of both worlds. It is the precision and quality of this German test rig that enables every customer to experience the quality of the Spanish design. Óscar Fernández: "We want to impress even the most demanding customers."

ES **CARGO**
MEISTERBOCK
EN SEAT DESDE
2001

Una vez más, volvemos al «Meisterbock». Este dispositivo de medición de apariencia sencilla, pero que en realidad es altamente inteligente y extremadamente preciso, es algo así como el centro neurálgico de la producción en Martorell. Los responsables técnicos, como Óscar Fernández, lo llaman el «modelo de referencia base». Aquí es donde cada una de las piezas de los talleres de prensas y de carrocería de SEAT en Martorell y Barcelona, y cada una de las piezas de plástico suministradas por los proveedores, deben ajustarse con una precisión de centésimas de milímetro.

El Meisterbock hace patente lo bien que SEAT combina lo mejor de dos mundos. Es la calidad y precisión del banco de pruebas alemán lo que permite que todos los clientes experimenten la calidad del diseño español. Óscar Fernández: «Queremos impresionar incluso a los clientes más exigentes».

DE **POSITION**
MEISTERBOCK
BEI SEAT SEIT
2001

Und wieder der Meisterbock: Die scheinbar so einfache, im Detail aber hochintelligente und extrem präzise Messanlage ist so etwas wie das Nervenzentrum von Entwicklung und Produktion in Martorell. Die verantwortlichen Ingenieure wie Óscar Fernández nennen es ein „Null-Referenz-Modell". Hier muss jedes Blechteil aus dem SEAT Presswerk und Karosseriebau in Martorell und Barcelona und auch jedes Kunststoffteil von einem Lieferanten auf den Hundertstelmillimeter passen.

Damit beweist der Meisterbock, wie sehr SEAT das Beste aus zwei Welten vereint: Denn erst die Qualität und Präzision dieses deutschen Prüfstands macht die Qualität des spanischen Designs für jeden Kunden erlebbar. Óscar Fernández: „Es geht uns darum, auch den anspruchsvollsten Kunden zu begeistern."

LIGHTING TECHNOLOGIES DEVELOPMENT
AT SEAT SINCE
1998

Carlos Elvira loves the light – when he's riding around on his quad bike in the bright Catalan sunshine on his days off. But he also loves the bright gleam of the LEDs when he's working in the Centro Tecnico on a headlamp for the next SEAT model. At 5,300 Kelvin, the colour temperature of LED headlamps is very similar to that of daylight, which puts less strain on the eyes than conventional illumination. "But light is not just about safety, it's also a matter of design," says Carlos Elvira. "We bring both together with passion." The light signature is an important characteristic of the brand. And, with its clear triangular form in the LED daytime driving lights and the rear lights, SEAT is once again unmistakeable.

ES CARGO
DESARROLLO DE TECNOLOGÍAS DE ILUMINACIÓN
EN SEAT DESDE
1998

A Carlos Elvira le encanta la luz. La disfruta al máximo cuando viaja en Quad con el brillo del atardecer como telón de fondo en su tiempo libre. Pero también le gusta el brillo de los LEDs cuando trabaja en el Centro Técnico en el desarrollo de los faros del próximo modelo de SEAT. Una temperatura de color de 5.300 Kelvin en los faros LED es muy similar a la de la luz solar, lo cual produce menos fatiga ocular que la iluminación convencional. «La luz no es solo una cuestión de seguridad, es también un tema de diseño», dice Carlos Elvira. «Combinamos ambos aspectos con auténtica pasión». La forma de la luz también es una característica importante de la marca. La clara forma triangular de la luz día con LEDs en los faros y en los pilotos posteriores hace que un SEAT sea realmente inconfundible.

DE POSITION
ENTWICKLUNG LICHTTECHNOLOGIE
BEI SEAT SEIT
1998

Carlos Elvira liebt das Licht. Die klare katalanische Sonne, wenn er in der Freizeit mit seinem Quad unterwegs ist. Aber auch das helle Strahlen der Leuchtdioden, wenn er im Centro Tecnico an einem Scheinwerfer für das nächste SEAT Modell arbeitet. Schließlich ist die Farbtemperatur der LED-Scheinwerfer mit 5.300 Kelvin dem Tageslicht sehr ähnlich, was die Augen weniger ermüdet als konventionelle Beleuchtung. „Licht ist aber nicht nur Sicherheit, Licht ist auch Design", sagt Carlos Elvira. „Wir verbinden beides mit Leidenschaft." Die Lichtsignatur ist ein wichtiges Erkennungszeichen der Marke. Und da ist SEAT mit seinem klaren Dreieck im LED-Tagfahrlicht und in den Heckleuchten wieder mal unverwechselbar.

SEAT AIMS TO BE NUMBER ONE WHEN IT COMES TO INNOVATIONS IN LIGHTING TECHNOLOGY.

ES Cuando se trata de innovación en tecnología de iluminación, SEAT quiere ser el número uno.

DE Bei Innovationen in der Lichttechnologie soll SEAT die Nummer eins werden.

LIGHTING TECHNOLOGIES DEVELOPMENT

CARLOS ELVIRA

PABLO BARRIOS

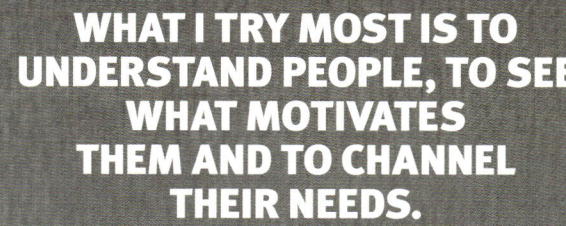

WHAT I TRY MOST IS TO UNDERSTAND PEOPLE, TO SEE WHAT MOTIVATES THEM AND TO CHANNEL THEIR NEEDS.

ES Me centro en entender a las personas, comprender su motivación y acertar en sus necesidades.
DE Mein höchstes Ziel ist es, die Menschen zu verstehen und zu erkennen, was sie motiviert und wo ihre Bedürfnisse sind.

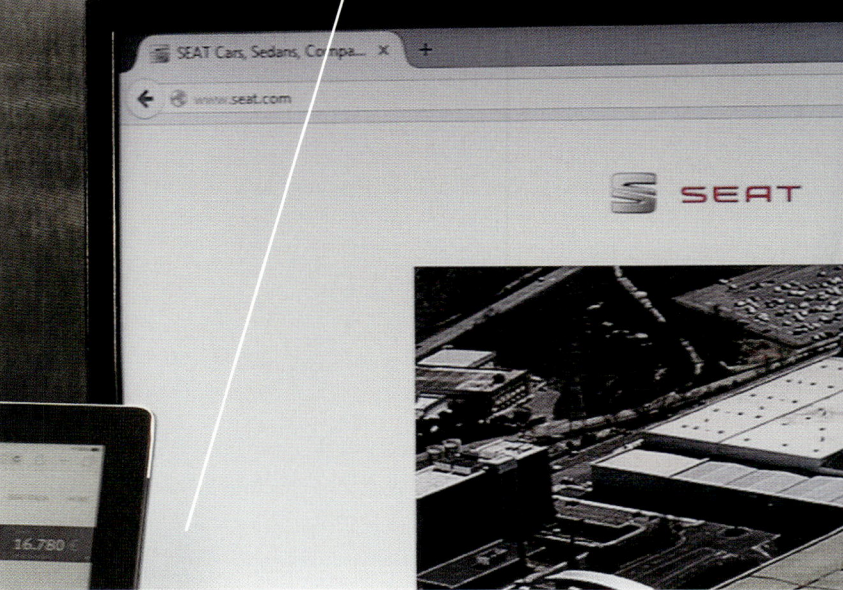

EN **POSITION**
DIGITAL MARKETING
AT SEAT SINCE
2012

Pablo Barrios is a big technology fan. He loves innovative gimmicks, makes extensive use of his smartphone and is very much at home in the networked world. This puts Pablo in exactly the same boat as many SEAT customers. And this is also why he is convinced that SEAT has to be right at the front of the pack in the use of digital technologies. The best possible in-car connectivity is a given for him, but just as important are the digital platforms that connect the SEAT brand to its customers.

"95 percent of car buyers find their information on the internet before making their first visit to a dealership," says Pablo. "And in very many cases, they have already made their decision. We therefore have to develop the right online tools to provide customers with information and advice quickly and directly. And we'll offer completely new services." Pablo is absolutely convinced that the digital world is a decisive factor for the future of the brand.

ES **CARGO**
MARKETING DIGITAL
EN SEAT DESDE
2012

Pablo Barrios es un gran fan de la tecnología. Le encantan los dispositivos innovadores, utiliza su smartphone para todo, y conectado a la red se siente como pez en agua. En este sentido, Pablo está en sintonía con muchos clientes de SEAT, y por eso precisamente está convencido de que SEAT debe estar a la vanguardia en el uso de tecnologías digitales. Una conectividad óptima en el coche es muy importante, y de igual forma lo es el uso de las plataformas digitales para la relación de la marca con sus clientes.

«El 95 por ciento de los compradores de automóviles se informan por Internet antes de visitar un concesionario por primera vez», dice Pablo. «Y en muchos casos nos visitan con la decisión ya tomada. Por lo tanto, debemos poner a disposición los recursos on-line para informarles y asesorarles de forma rápida y directa. De esta manera podremos ofrecer productos y servicios totalmente innovadores». Pablo está convencido de que el mundo digital es decisivo para el futuro de la marca.

DE **POSITION**
DIGITALES MARKETING
BEI SEAT SEIT
2012

Pablo Barrios ist ein großer Technik-Fan. Er liebt innovative Gimmicks, nutzt sein Smartphone intensiv und ist in der vernetzten Welt zu Hause. Damit tickt Pablo exakt genauso wie viele der Kunden von SEAT. Und deshalb, so ist er überzeugt, muss SEAT bei der Nutzung digitaler Technologien auch ganz vorne sein. Die optimale Connectivity im Auto ist für ihn selbstverständlich, ebenso wichtig sind aber digitale Plattformen für die Verbindung der Marke SEAT zu ihren Kunden.

„95 Prozent der Automobilkäufer informieren sich im Internet, bevor sie das erste Mal einen Händlerbetrieb besuchen", weiß Pablo. „Und in sehr vielen Fällen ist dann die Entscheidung schon gefallen. Also müssen wir die richtigen Online-Tools entwickeln, um die Kunden schnell und direkt zu informieren und zu beraten. Und wir werden ganz neue Services und Dienstleistungen anbieten." Denn für die Zukunft der Marke ist die digitale Welt entscheidend, davon ist Pablo überzeugt.

EN   POSITION
DESIGN LEADER EXTERIOR COMPONENTS & CUPRA
AT SEAT SINCE
2009

The design is often the decisive motivation for people when choosing a car. SEAT Design is spot on here. "This is why we design vehicles that combine a perfect form with a high degree of functionality – unique cars with a distinctive personality." Tony Gallardo works in SEAT Design on vehicle exteriors and, as Design Leader, is responsible for the powerful look of the CUPRA models. "We want to inspire our customers. We would like them to be proud of their vehicle."

SEAT Design achieves this with creativity, with passion and, of course, with the brand's classic attention to the tiniest detail. Tony Gallardo: "We create true value for our customers with our ideas. Enjoying the day-to-day work interacting with other departments and making the most of every opportunity for improvement are key factors in achieving our project objectives."

ES   CARGO
RESPONSABLE DE DISEÑO DE COMPONENTES EXTERIORES
Y MODELOS CUPRA
EN SEAT DESDE
2009

A la hora de elegir un coche, el diseño es ciertamente uno de los factores concluyentes. En este aspecto, el Área de Diseño de SEAT da en el blanco. «Diseñamos coches que combinan un alto nivel de funcionalidad con formas perfectas, coches únicos dotados de personalidad propia». Tony Gallardo trabaja en el departamento de Diseño de SEAT, concretamente en componentes exteriores y, como jefe de diseño, es responsable del potente aspecto de los modelos CUPRA. «Queremos ilusionar a nuestros clientes. Nos gusta que se sientan orgullosos de su coche».

Diseño SEAT lo consigue con creatividad y pasión, y también gracias a la especial atención a los pequeños detalles que caracteriza a la marca. Tony afirma: «Nuestras ideas se traducen en valor real para nuestros clientes». Disfrutar del trabajo día a día interactuando con otras áreas, aprovechando cada oportunidad para mejorarlo, es una de las claves para superar los objetivos establecidos al inicio de los proyectos.

DE   POSITION
LEITENDER DESIGNER EXTERIEUR-KOMPONENTEN UND CUPRA MODELLE
BEI SEAT SEIT
2009

Das Design liefert sehr oft die entscheidende Motivation, warum sich Menschen für ein Automobil entscheiden. SEAT Design ist hier auf einem perfekten Weg. „Deshalb entwerfen wir Fahrzeuge, die eine perfekte Form mit hoher Funktionalität verbinden. Einzigartige Automobile mit einer eigenen Persönlichkeit." Tony Gallardo arbeitet bei SEAT Design am Exterieur der Fahrzeuge und ist als leitender Designer für den kraftvollen Auftritt der CUPRA-Modelle verantwortlich. „Wir wollen unsere Kunden inspirieren. Wir möchten, dass sie stolz sind auf ihr Fahrzeug."

SEAT Design gelingt das mit Kreativität, mit Leidenschaft und natürlich mit der für die Marke typischen Liebe auch zum kleinsten Detail. Tony Gallardo: „Mit unseren Ideen schaffen wir einen echten Wert für unsere Kunden. Wir genießen die tägliche Zusammenarbeit mit den anderen Bereichen und wir nutzen jede Gelegenheit, um besser zu werden – das sind die entscheidenden Elemente um unsere Ziele zu erreichen."

WE CREATE TRUE VALUE FOR OUR CUSTOMERS WITH OUR IDEAS.

ES Nuestras ideas se traducen en valor real para nuestros clientes.
DE Mit unseren Ideen schaffen wir einen echten Wert für unsere Kunden.

DESIGN LEADER EXTERIOR COMPONENTS & CUPRA

TONY GALLARDO

JOAN ROIG

DYNAMIC PERFORMANCE MEANS DRIVING FUN. AND OUR DYNAMIC PERFORMANCE COMES FROM PASSION.

ES El comportamiento dinámico es sinónimo de disfrute al volante. Y nuestro comportamiento dinámico es fruto de la pasión.
DE Dynamik bedeutet Fahrspaß. Unsere Dynamik entsteht aus Leidenschaft.

EN POSITION
CHASSIS DEVELOPMENT
AT SEAT SINCE
1998

Joan Roig is a lover of precision and elegance – the precision of sporty driving and the elegance of agile and light-footed movement. Joan Roig and his colleagues are responsible for an important part of the SEAT DNA – for dynamic driving fun.

Only the perfectly tuned interaction of steering, suspension, springs, dampers, brakes, wheels, tyres and a host of other parts creates that special SEAT feeling behind the wheel. Systematic tuning, patient refinement and good driving skills are as important for a chassis engineer as a precise idea of how a SEAT has to feel. Joan Roig: "It's the passion for the perfect feel that drives us."

ES CARGO
DESARROLLO CHASIS
EN SEAT DESDE
1998

Joan Roig es un amante de la precisión y la elegancia. La precisión de la deportividad en la conducción y la elegancia de los movimientos ágiles y dinámicos. Joan Roig y sus compañeros son responsables de una parte importante del ADN de SEAT: El disfrute de una conducción dinámica.

Solo la interacción en perfecta sintonía de, dirección, suspensión, muelles, amortiguadores, frenos, ruedas, neumáticos y otras muchas partes crea la especial sensación que genera SEAT detrás del volante. El ajuste sistemático, un perfeccionamiento sosegado y buenas habilidades de conducción, son tan importantes para un ingeniero de chasis como la idea exacta de cómo debe sentirse un SEAT. Joan Roig: «Nos mueve la pasión por buscar la sensación perfecta»

DE POSITION
ENTWICKLUNG FAHRWERK
BEI SEAT SEIT
1998

Joan Roig liebt Präzision und Eleganz – die Präzision des sportlichen Fahrens und die Eleganz einer leichtfüßigen Bewegung. Gemeinsam mit seinen Kollegen ist Joan Roig für einen wichtigen Teil der DNA von SEAT verantwortlich: den dynamischen Fahrspaß.

Erst das perfekt abgestimmte Zusammenspiel von Lenkung, Federung, Dämpfung, Bremsen, Rädern, Reifen und vieler weiterer Bauteile schafft das besondere SEAT Fahrgefühl. Die konsequente Abstimmung, das geduldige Verfeinern und herausragendes Fahrkönnen sind für einen Fahrwerksingenieur ebenso wichtig wie die genaue Vorstellung davon, wie sich ein SEAT anfühlen muss. Joan Roig: „Es ist die Passion für das perfekte Fahrgefühl, die uns antreibt."

SEAT
EMOTIONAL TECH-NOLOGY
HIGH-TECH FOR DRIVING PLEASURE WITH RESPONSIBILITY

INNOVATIVE LIGHTING TECHNOLOGY
HIGH LIGHT

FULL LED

EN SEAT led the way when it introduced Full-LED technology into the segment. The LED headlamps in the Leon and Toledo illuminate the road at a colour temperature of 5,300 Kelvin. Because this is very similar to that of daylight, it puts very little strain on the eyes. LEDs shine brightly in terms of efficiency, too, with minimum energy consumption.

ES SEAT fue pionera en el sector al introducir la tecnología Full LED. Los faros LED del León y del Toledo iluminan la carretera con una temperatura de color de 5.300 grados Kelvin. Se trata de una luz muy similar a la solar, por lo que produce muy poca fatiga ocular. Los LED también brillan por su eficiencia, ya que el consumo de energía es mínimo.

DE SEAT brachte als Pionier die Voll-LED-Technologie ins Segment. Die LED-Scheinwerfer im Leon und Toledo leuchten die Straße mit 5.300 Kelvin Farbtemperatur aus. Da dies dem Tageslicht sehr ähnlich ist, lässt es die Augen kaum ermüden. In punkto Effizienz brillieren die Leuchtdioden durch minimale Energieaufnahme.

SIGNATURE DESIGN

EN Headlamps and tail lights are part of the characteristic SEAT design DNA. The light signature of the LED daytime running lights and LED tail lights is unique and gives each and every SEAT a determined and powerful expression.

ES Los faros y las luces traseras forman parte del ADN característico del diseño de SEAT. La forma de las luces diurnas y traseras de LED es única, y confiere una expresión potente y decidida a todos y cada uno de los modelos de SEAT.

DE Scheinwerfer und Heckleuchten sind Teil der charakterstarken SEAT Design-DNA: Die Lichtsignatur im LED-Tagfahrlicht und in den LED-Heckleuchten ist einzigartig und gibt jedem SEAT ein entschlossenen und kraftvollen Ausdruck.

HIGH QUALITY INTERIOR

INNER BEAUTY

Full Link

EASY CONNECT

EN Operation is functional and intuitive – touchscreen with proximity sensor, logically arranged buttons on the steering wheel, good ergonomics for relaxed long-distance travel.

ES El manejo es funcional e intuitivo: pantalla táctil con sensor de proximidad, botones organizados de forma lógica en el volante y una excelente ergonomía para que los viajes largos sean cómodos.

DE Die Bedienung ist funktional und intuitiv: Touchscreen mit Näherungssensor, logisch sortierte Tasten im Lenkrad, gute Ergonomie für entspanntes Reisen auf langen Strecken.

DRIVER ORIENTED

EN The driver is the focal point. An important part of a precise driving feel is clear and precisely laid out instruments – full information with full concentration on the road.

ES El centro de atención está en el conductor. Experimentar una conducción precisa depende, en cierta medida, de que los instrumentos estén dispuestos de forma clara y precisa. Poder contar con toda la información necesaria ayuda a concentrarse totalmente en la carretera.

DE Der Fahrer steht im Mittelpunkt. Zu einem präzisen Fahrgefühl gehören klar und präzise gezeichnete Instrumente. Volle Information bei voller Konzentration auf die Straße.

HIGH QUALITY

EN Premium surfaces, exact tolerances, switches and buttons with a precise click, finely stitched leather – a SEAT is a joy to touch, every single day.

ES Superficies de primera calidad, exactitud extrema, interruptores y botones precisos, cuero delicadamente bordado... Tocar un SEAT siempre es un placer.

DE Hochwertige Oberflächen, exakte Passungen, Schalter und Knöpfe mit einem präzisen Klick-Gefühl, fein vernähtes Leder. Einen SEAT zu berühren macht Freude. Jeden Tag.

MIRROR-LINK

EN Convenient and safe – operate your smartphone with the SEAT touchscreen. FullLink enables this functionality for Android or Apple iOS devices.

ES Utiliza el teléfono móvil de forma cómoda y segura gracias a la pantalla táctil de SEAT. La tecnología FullLink es compatible con dispositivos Android y dispositivos iOS de Apple.

DE Komfortabel und sicher – mit dem Touchscreen im SEAT das Smartphone bedienen. Dank FullLink für Geräte mit Android oder Apple iOS.

SEAT CONNECT APP

EN Draw, don't write. Speak, don't type. The SEAT ConnectApp offers exclusive features like gesture control and speech-to-text.

ES No escribas, señala. No teclees, habla. ConnectApp de SEAT dispone de funciones exclusivas como el control mediante gestos y el reconocimiento de voz.

DE Zeichnen statt schreiben, sprechen statt tippen: Die SEAT ConnectApp bietet exklusive Features wie Gestensteuerung und Speech-to-Text.

POWERED BY SAMSUNG

EN The exclusive cooperation keeps SEAT ahead of the competition. Together with Samsung, SEAT is developing innovative solutions for mobile connectivity.

ES La colaboración exclusiva entre SEAT y Samsung nos coloca por delante de la competencia. Juntos estamos desarrollando innovadoras soluciones de conectividad móvil.

DE Die exklusive Kooperation sichert den Vorsprung: Zusammen mit Samsung entwickelt SEAT innovative Lösungen für mobile Connectivity.

DYNAMIC EFFICIENCY
ECO DRIVE

EN Dynamic driving fun with responsibility – in terms of fleet consumption, SEAT is one of the world's most efficient automotive brands. Between 2006 and 2014 alone, SEAT reduced CO_2 emissions by 21 percent. Low weight, good aerodynamics and optimised drives deliver more kilometres for every litre of fuel.

ES Diversión garantizada por una conducción dinámica pero responsable. En cuanto a consumo, SEAT es una de las marcas de automóviles más eficientes del mundo. Entre 2006 y 2014, SEAT redujo las emisiones de CO_2 en un 21 por ciento. Fabricamos coches ligeros, con una buena aerodinámica y una transmisión optimizada, lo que se traduce en más kilómetros por cada litro de combustible.

DE Dynamischer Fahrspaß mit Verantwortung: SEAT zählt im Flottenverbrauch zu den effizientesten Automobilmarken der Welt. Allein von 2006 bis 2014 hat SEAT die CO_2-Emissionen um 21 Prozent gesenkt. Geringes Gewicht, gute Aerodynamik und optimierte Triebwerke machen mehr Kilometer aus jedem Liter Kraftstoff.

ECO MOTIVE

EN 85 kW / 115 PS, forceful torque, but fuel consumption of just 4.3 litres – the Leon 1.0 TSI ECOMOTIVE is compelling for its sporty efficiency.

ES Con un efectivo par motor de 85 kW / 115 CV y un consumo de tan solo 4,3 litros, el León 1.0 TSI ECOMOTIVE convence por su eficiencia deportiva.

DE 85 kW / 115 PS, kräftiges Drehmoment, aber nur 4,3 Liter Verbrauch – der Leon 1.0 TSI ECOMOTIVE begeistert durch sportliche Effizienz.

NATURAL GAS

EN SEAT's CNG vehicles already use an alternative fuel that is low in cost and low in emissions.

ES Los coches con motor GNC de SEAT ya utilizan un combustible alternativo de bajo coste que produce menos emisiones.

DE Die Erdgasfahrzeuge von SEAT nutzen bereits heute einen alternativen Kraftstoff mit geringen Kosten und günstigen Emissionen.

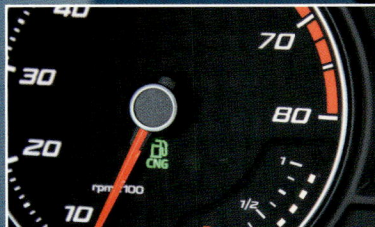

TAKE THREE

EN The new generation of three-cylinder turbocharged TSI and TDI engines is impressive for responsive performance and low fuel consumption.

ES La nueva generación de motores turbo de tres cilindros TSI y TDI impresiona por su rendimiento y por su bajo consumo de combustible.

DE Die neue Generation von Dreizylinder-TSI- und TDI-Motoren mit Turboaufladung begeistert durch spontane Leistung und niedrigen Verbrauch.

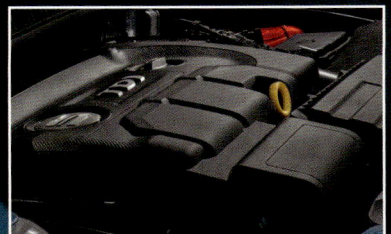

ELECTRIC POWER

EN SEAT is developing models with plug-in hybrid and battery-electric drive. They will be offered in the line-up as soon as the market conditions are right.

ES SEAT está desarrollando modelos de vehículos eléctricos de batería e híbridos enchufables. Saldrán a la venta cuando las condiciones del mercado lo permitan.

DE SEAT entwickelt Modelle mit Plug-in-Hybrid und reinen Elektroantrieb. Sie werden angeboten, sobald die Marktbedingungen stimmen.

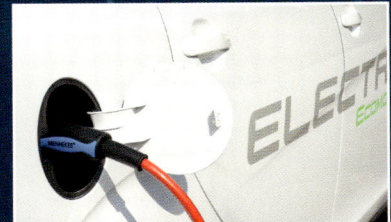

THE UNIQUE
CUPRA FORMULA

En CUPRA is dynamic driving fun in its purest form. CUPRA is full-on performance and outstanding technology, but also compelling efficiency and perfect everyday usability – the Leon CUPRA is the perfect embodiment of SEAT brand values. CUPRA is SEAT.

ES La fórmula CUPRA ofrece la diversión de la conducción dinámica en estado puro, el rendimiento máximo y una tecnología espectacular, pero también convence con su eficiencia y su absoluta facilidad. El León CUPRA personifica a la perfección los valores de la marca SEAT. CUPRA es SEAT.

DE CUPRA ist dynamischer Fahrspaß pur. CUPRA ist volle Performance und überlegene Technologie, aber auch überzeugende Effizienz und perfekte Alltagstauglichkeit – der Leon CUPRA verkörpert die Markenwerte von SEAT in perfekter Weise. CUPRA ist SEAT.

1

DYNAMIC CHASSIS CONTROL

EN The dynamic chassis control (DCC) has been fine-tuned specifically for the CUPRA and is now even faster and more sensitive than ever, adapting the chassis characteristics to the prevailing conditions in a matter of milliseconds.

ES Hemos ajustado el control de chasis adaptativo (DCC) específicamente para el CUPRA, por lo que ahora es más rápido y sensible que nunca. Es capaz de adaptar las características del chasis a las condiciones predominantes en cuestión de milésimas de segundo.

DE Die speziell für den CUPRA perfektionierte adaptive Fahrwerksregelung (DCC) ist nun noch sensibler sowie schneller und passt die Charakteristik des Fahrwerks binnen Millisekunden den aktuellen Bedingungen an.

PROGRESSIVE STEERING

EN The hi-tech steering system adapts to road speed for even greater handling precision, making frequent grip changes a thing of the past. The SEAT Leon CUPRA boasts incredibly dynamic and stunningly precise handling.

ES El sistema de dirección de alta tecnología se adapta a la velocidad en carretera para que la precisión de manejo sea aún mayor, haciendo que los cambios frecuentes de adherencia sean cosa del pasado. El SEAT León CUPRA ofrece un manejo que sorprende por su dinamismo y precisión.

DE Die Hightech-Lenkung passt sich der Fahrgeschwindigkeit an und macht das Handling noch präziser. So gehört häufiges Umgreifen der Vergangenheit an: Der SEAT Leon CUPRA überzeugt mit seinem dynamischen und seinem faszinierend präzisen Handling.

③ # CUPRA DRIVE PROFILE

EN As soon as the CUPRA Drive Profile is activated, the shifting of the DSG becomes even sportier, the gas pedal even more responsive and the engine sound more powerful. And for those more peaceful days in life, there is the Comfort Drive Profile – at the touch of a button.

ES En cuanto se activa el CUPRA Drive Profile, los cambios de la caja DSG se vuelven más deportivos, el acelerador responde aún más y el sonido del motor es más potente. Y para los días tranquilos, basta con pulsar un botón para cambiar al Comfort Drive Profile.

DE Ist das CUPRA Drive Profile erst einmal aktiviert, schaltet das DSG-Getriebe noch sportlicher, reagiert das Gaspedal noch sensibler und klingt der Motor noch kraftvoller. Und für die ruhigeren Tage des Lebens gibt es das Drive Profile Comfort. Auf Knopfdruck.

DIFFERENTIAL LOCK

EN The front-axle differential lock lifts the traction and handling of the SEAT Leon CUPRA to a previously unheard-of level. In extreme cases, all the torque can be sent to one wheel, which improves handling and prevents understeer.

ES El bloqueo del diferencial del eje delantero lleva la tracción y el manejo del SEAT León CUPRA a niveles nunca vistos. En casos extremos es posible llevar todo el par motor a una rueda, lo que mejora el manejo y evita el subviraje.

DE Die Vorderachs-Differenzialsperre optimiert Traktion und Fahrdynamik des SEAT Leon CUPRA auf nie dagewesene Weise. Im Extremfall kann das gesamte Antriebsmoment auf ein Rad geleitet werden, das verbessert das Handling und verhindert Untersteuern.

POWERTRAIN MASTERPIECE

HOT HEART

EN The 2.0 TSI in the Leon CUPRA is a true sports engine. It responds immediately to every movement of the gas pedal, and thrills with high-revving agility – accompanied by a rich and refined sound. It comes with an incredible package of hi-tech with ground-breaking features, including a dual injection system, whereby the petrol direct injection is complemented under partial load by manifold injection. The turbocharger is incredibly responsive.

ES El 2.0 TSI del León CUPRA es un motor realmente deportivo. Responde inmediatamente a cada movimiento del acelerador y emociona gracias a su agilidad en las altas revoluciones, siempre con un sonido rico y refinado. Incluye un increíble paquete de alta tecnología con funciones revolucionarias, como un sistema de doble inyección mediante el que la inyección directa de gasolina se complementa con la inyección por tubo de aspiración durante el funcionamiento con carga parcial. La rapidez de respuesta del turbocompresor es increíble.

DE Der 2.0 TSI im Leon CUPRA ist ein echter Sportmotor. Er antwortet spontan auf jede Bewegung des Gaspedals und begeistert durch leichtfüßige Drehfreude – begleitet von einem sonoren und souveränen Sound. Dazu gehört ein absolutes Hightech-Paket mit wegweisenden Features: Beim dualen Einspritzsystem wird die Benzin-Direkteinspritzung im Teillastbereich durch eine Saugrohreinspritzung ergänzt, der Turbolader spricht besonders schnell an.

TOP PER-FORMANCE INCLUDED

EN CUPRA stands for power, performance, handling – and the new Leon CUPRA fulfils this promise perfectly. The CUPRA 290 with DSG transmission catapults itself to 100 km/h in just 5.7 seconds. Relative to this, the Leon CUPRA is astonishingly fuel efficient.

ES CUPRA es sinónimo de potencia, rendimiento y facilidad de manejo, y el nuevo León se ajusta perfectamente a esta definición. El CUPRA 290 con transmisión DSG llega a los 100 km/h en solo 5,7 segundos. Lo que resulta aún más sorprendente es que el León CUPRA consume muy poco combustible.

DE CUPRA steht für Leistung, Performance, Fahrdynamik – und der neue Leon CUPRA erfüllt diese Versprechen in perfekter Weise: Der CUPRA 290 mit DSG-Getriebe katapultiert sich in gerade mal 5,7 Sekunden auf Tempo 100. Gemessen daran ist der Leon CUPRA überraschend effizient.

PERFECT FILL

EN The cylinder fill of the 2.0 TSI is always spot on – the camshafts are adjustable, with the Valve Lift system regulating valve travel. The exhaust manifold is integrated into the cylinder head and is part of the thermal management. A new kind of coating on the aluminium pistons and bearings keeps friction to a minimum.

ES La eficiencia volumétrica del 2.0 TSI siempre es acertada, la mezcla perfecta. Los árboles de levas son regulables y el sistema de elevación de las válvulas de admisión controla la carrera de las válvulas. El colector de escape está integrado en la culata del cilindro y forma parte del sistema de gestión térmica. El uso de un nuevo tipo de revestimiento en los pistones de aluminio y los cojinetes permite reducir la fricción interna al mínimo.

DE Die Zylinder des 2.0 TSI werden stets optimal gefüllt: Die Nockenwellen sind verstellbar, das Valve Lift-System reguliert den Hub der Ventile. Der Abgaskrümmer ist in den Zylinderkopf integriert und Teil des intelligenten Thermomanagements. Eine neuartige Beschichtung der Aluminiumkolben und der Lager hält die Reibung gering.

SEAT
THE
HISTORY
WHERE WE COME FROM

1957
SEAT 600
MAKING
SPAIN MOBILE

EN Launched in 1957, the 600 changed both Spanish industry and Spanish lives. It ignited the passion for four wheels and got the whole country moving. Seeing one of these on the streets today is seeing a real little hero in action.

ES En 1957 se lanzó al mercado el 600, lo que supuso un cambio para el sector en España, pero también para las vidas de los españoles. Encendió la pasión por las cuatro ruedas e hizo que todo el país empezase a desplazarse. Cuando vea uno de estos modelos en las calles, piense que está ante un pequeño héroe en acción.

DE Vorgestellt im Jahr 1957, verwandelte der SEAT 600 sowohl die spanische Industrie als auch das Leben der Spanier. Er entzündete die Leidenschaft für vier Räder und brachte das gesamte Land in Bewegung. Wenn man heute einen 600 auf der Straße sieht, dann sieht man einen wahren, kleinen Helden.

1984
SEAT IBIZA
CREATING
AN OWN IDENTITY

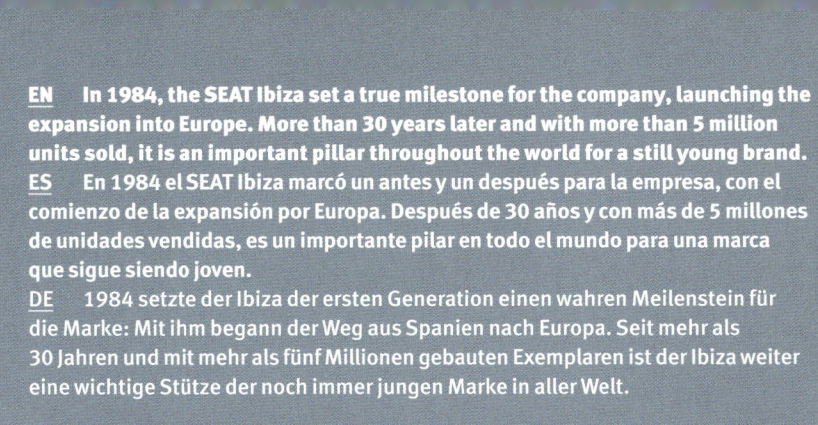

EN In 1984, the SEAT Ibiza set a true milestone for the company, launching the expansion into Europe. More than 30 years later and with more than 5 million units sold, it is an important pillar throughout the world for a still young brand.

ES En 1984 el SEAT Ibiza marcó un antes y un después para la empresa, con el comienzo de la expansión por Europa. Después de 30 años y con más de 5 millones de unidades vendidas, es un importante pilar en todo el mundo para una marca que sigue siendo joven.

DE 1984 setzte der Ibiza der ersten Generation einen wahren Meilenstein für die Marke: Mit ihm begann der Weg aus Spanien nach Europa. Seit mehr als 30 Jahren und mit mehr als fünf Millionen gebauten Exemplaren ist der Ibiza weiter eine wichtige Stütze der noch immer jungen Marke in aller Welt.

2012 SEAT LEON
BECOMING A POWER BRAND

EN A new era begins. With its unique "Leon Formula", the current Leon places the brand on a whole new level. Design and functionality; performance and comfort; technology and quality – all of it, of course, conceived and brought to life with SEAT's unique emotion.

ES Empieza una nueva era. Con su exclusiva «Fórmula León», el nuevo León sitúa a la marca en un nivel completamente nuevo. Diseño y funcionalidad, rendimiento y confort, tecnología y calidad, todo ello diseñado y materializado con la emoción incomparable de SEAT.

DE Eine neue Ära beginnt: Mit seiner einzigartigen „Leon-Formel" bringt der aktuelle Leon die Marke auf eine neue Stufe. Design und Funktionalität, Dynamik und Komfort, Technologie und Qualität – natürlich alles konzipiert und umgesetzt mit der einzigartigen Emotion von SEAT.

SEAT CHALLENGE IS PART OF OUR DNA

SEAT LEON CUP RACER

EN Performance is part of the SEAT brand. Our rolling test lab for maximum driving fun is the Leon CUP RACER. Not only is it successful on the race track in SEAT customer sport, its genes are also evident in every road-going SEAT CUPRA model. At SEAT, motorsport is never an end in itself, but part of the work to develop and perfect our automobiles.

ES El rendimiento forma parte de la marca SEAT. El León CUP RACER es nuestro laboratorio de pruebas para obtener la máxima diversión de conducción. Además de gozar de un gran éxito en las pistas de carreras entre los clientes deportivos de SEAT, sus genes también son patentes en los modelos para carretera de SEAT CUPRA. En SEAT, el automovilismo deportivo no es un fin en sí mismo sino parte del trabajo que hacemos para desarrollar y perfeccionar nuestros coches.

DE Dynamik gehört zur Marke SEAT. Unser rollendes Testlabor für maximalen Fahrspaß ist der Leon CUP RACER. Er ist nicht nur erfolgreich im SEAT Kundensport auf den Rennstrecken, seine Gene sind auch spürbar in jedem SEAT CUPRA-Modell auf der Straße. Denn Motorsport ist bei SEAT niemals Selbstzweck, sondern Teil der Entwicklungsarbeit und Perfektionierung unserer Automobile.

LEON CUP RACER

OUR LAB FOR MAXIMUM PERFORMANCE

YOU ARE IN THE DRIVING SEAT

CFRP BUCKET SEAT

EN The best possible safety from a high-strength steel cage and CFRP bucket seat with special protection for head and neck.

ES La jaula de acero de alta resistencia y el asiento de plástico reforzado con fibra de carbono, que cuenta con una protección especial para la cabeza y el cuello, garantizan la máxima seguridad posible.

DE Optimale Sicherheit durch höchstfesten Stahlkäfig und CFK-Schalensitz mit speziellem Schutz für Kopf und Nacken.

MULTIFUNCTION STEERING WHEEL

EN The multifunction steering wheel and TFT info display mean the driver is always in full control of his racing machine.

ES Gracias al volante multifunción y la pantalla TFT de información, el conductor siempre tiene el control absoluto de esta máquina de competición.

DE Mit Multifunktionslenkrad und TFT-Infodisplay behält der Pilot stets die Kontrolle über sein Renngerät.

SHIFT LEVER

EN Shift lever for the sequential 6-speed race transmission.

ES Palanca de cambios de la transmisión secuencial de 6 velocidades.

DE Schalthebel für sequentielles 6-Gang-Renngetriebe.

THE FASCINATION OF SPEED

A PRECISION MACHINE FOR PHENOMENAL FUN TANGIBLE IN EVERY SEAT

EN A design that conveys concentrated power and yet is very clearly a Leon – the SEAT design DNA works on the race track, too. The CUP RACER commands enormous potential in tough head-to-head competition, with its two-litre turbocharged engine, 243 kW / 330 PS, front-wheel drive and differential lock. And every series-production model benefits from the development work done by SEAT Sport, not just the supreme performance of the Leon CUPRA.

ES Un diseño que transmite una potencia concentrada sin dejar de ser un León, porque el ADN del diseño de SEAT también funciona en la pista de carreras. El CUP RACER demuestra tener un gran potencial en las competiciones más reñidas, con su motor turbo de dos litros de 243 kW / 330 CV, tracción delantera y bloqueo del diferencial. No se trata únicamente del extraordinario rendimiento que ofrece el León CUPRA, todos los modelos de producción de serie se benefician del trabajo de desarrollo realizado por SEAT Sport.

DE Ein Design, das geballte Kraft ausdrückt, und doch ganz klar ein Leon: Die Design-DNA von SEAT passt auch auf die Rennstrecke. Mit seinem Zweiliter-Turbomotor, 243 kW / 330 PS, Frontantrieb und Sperrdifferenzial besitzt der CUP RACER hohes Potenzial im harten Wettbewerb. Doch auch jedes Serienmodell profitiert vom Entwicklungslabor SEAT Sport, nicht nur die überlegene Fahrdynamik des Leon CUPRA.

EN   Dr. Matthias Rabe,
       Board Member for Research
       and Development SEAT S.A.

ES   Dr. Matthias Rabe,
       Vicepresidente ejecutivo
       de Investigación y Desarrollo
       de SEAT S.A.

DE   Dr. Matthias Rabe,
       Mitglied des Vorstands
       für Forschung und Entwicklung
       SEAT S.A.

MATTHIAS RABE'S PERSPECTIVE

GERMAN ENGINEERING AND SPANISH EMOTION – THIS COMBINATION MAKES US UNIQUE.

ES La ingeniería alemana combinada con la emoción española,
una simbiosis que nos hace únicos.

DE German Engineering und spanische Emotion:
Diese Verbindung macht uns einzigartig.

WHAT DRIVES US IS THE PURSUIT OF PERFECTION.

ES Lo que nos motiva es la búsqueda de la perfección.
DE Was uns antreibt, ist der Wille zur Perfektion.

EN Dr. Matthias Rabe loves dynamic performance. Fast, but always controlled, precise and elegant movement – on skis, on water, on the road, of course, and even in the air. SEAT's R&D chief is not only a committed driver, he is also passionate about flying. He has been in the air since the age of 14, starting in the classic way with gliding, before adding an engine to the equation. Aerobatic flying was the big challenge. Today, Dr. Rabe takes every opportunity he has to fly across the stunning Catalonian landscape. What really drives Matthias Rabe about this third dimension is the pursuit of perfection. In many situations, flying does not tolerate errors, calling for the highest levels of concentration, the absolute control of man over machine. In the cockpit, you always have to be looking ahead, looking into the future, anticipating what is coming … "A pilot is really good when he is always a little bit ahead of the machine, when he always reckons with what can possibly come next."

EVERY SEAT POSSESSES A DISTINCT CHARACTER.

This is exactly what he instils every day into his team of more than 1,000 engineers. "We are working on the future, developing more than five years ahead, defining exciting, successful new models." This future will be guided by the Leon as SEAT "leonizes the brand". This means introducing the ingredients of the Leon success into forthcoming models. Design and functionality, dynamics and comfort, accessibility and quality – all combined with the best technology to guarantee emotion.

Matthias Rabe also needs fully focused concentration on the race track, another of his passions. Dr. Rabe was actively involved even in the fine-tuning and testing of the Leon CUP RACER and the "civilian" top-of-the-range model, the Leon CUPRA. "SEAT is seen as a very sporty brand," says the technical director, "and our CUPRA range lives up to that expectation."

Dr. Matthias Rabe is also well aware that driving the brand forward is achievable only with the best possible products. This obviously includes the very latest technologies, as SEAT naturally makes use of the Volkswagen Group's enormous knowledge pool. But what is more important is the distinct character of the car. And this is created in SEAT's very specific Spanish gene pool. "We are unique in combining German engineering and Spanish passion," says Dr. Rabe with conviction, pointing to the Leon ST CUPRA by way of example. Not only is it the most beautiful car in its class and the fastest on the Nürburgring, it is also the most fuel efficient on the road.

ES El Dr. Matthias Rabe es un admirador del rendimiento dinámico. Le gusta que los movimientos sean rápidos, pero siempre elegantes y controlados, ya sea esquiando, en el agua, en la carretera e incluso en el cielo. Además de ser un conductor consumado, el ingeniero jefe de SEAT también es un apasionado de los aviones. Lleva volando desde que tenía 14 años, cuando empezó a practicar con el clásico ala delta antes de añadir un motor a la ecuación. El vuelo acrobático fue para él un gran reto. Hoy, Rabe aprovecha cualquier oportunidad para surcar los cielos del impresionante paisaje catalán. Lo que realmente apasiona a Matthias Rabe de esta tercera dimensión es la búsqueda de la perfección. Son muchas las situaciones en las que no hay margen de error cuando se vuela, lo que requiere la máxima concentración, el control absoluto del hombre sobre la máquina. En la cabina siempre hay que estar alerta, pensar en el futuro inmediato, prever lo que ocurrirá… «Para ser realmente bueno, un piloto siempre debe ir un poco por delante de la máquina, siempre debe ser capaz de anticipar lo que va a pasar».

CADA SEAT TIENE SU PROPIO CARÁCTER.

Esto es justo lo que transmite cada día a su equipo de más de 1000 ingenieros. «Estamos preparando el futuro, desarrollando a más de cinco años vista, creando nuevos modelos de éxito que emocionen». Este futuro estará definido por el León, que impregnará todos los aspectos de la marca SEAT. Esto implica introducir los ingredientes del éxito del León en futuros modelos. Diseño y funcionalidad, dinámica y comodidad, accesibilidad y calidad… Y todo combinado con la mejor tecnología para dar lugar a productos emocionantes.

Matthias Rabe también tiene que estar plenamente concentrado en la pista de carreras, otra de sus pasiones. El Dr. Rabe participó de forma activa incluso en los procesos de ajuste y prueba del León CUP RACER y del modelo más alto de la gama para uso en carretera, el León CUPRA. «SEAT se considera una marca muy deportiva, y nuestra gama CUPRA está a la altura en este sentido», afirma el director técnico.

Matthias Rabe también es consciente de que la marca solo progresará si cuenta con los mejores productos. Esto incluye, por supuesto, las últimas tecnologías, ya que SEAT utiliza de forma natural el vasto conocimiento colectivo del Grupo Volkswagen. Pero lo que realmente importa es la marcada personalidad del coche, que emerge del acervo genético español tan propio de SEAT. «Somos únicos en combinar la ingeniería alemana con la pasión española», afirma el Dr. Rabe con convicción, y menciona el León ST CUPRA como ejemplo. Además de ser el coche más bonito de su categoría y el más rápido en Nürburgring, es el que consume menos combustible en la carretera.

DE Dr. Matthias Rabe liebt die Dynamik. Die schnelle, aber stets kontrollierte, präzise und elegante Bewegung – auf den Skiern, auf dem Wasser, natürlich auf der Straße, aber auch in der Luft. Der SEAT Vorstand für Forschung und Entwicklung ist nicht nur begeisterter Autofahrer, sondern ebenso passionierter Flieger. Seit seinem 14. Lebensjahr ist er in der Luft, ganz klassisch begonnen hat er als Segelflieger. Dann kam der Motor dazu, Kunstflug war die große Herausforderung. Heute nutzt Rabe jede Gelegenheit für Runden über die faszinierenden Landschaften Kataloniens. Was Matthias Rabe in dieser dritten Dimension wirklich antreibt, ist der Wille zur Perfektion. Das Fliegen verzeiht in vielen Situationen keinerlei Fehler, hier ist höchste Konzentration gefordert, die absolute Kontrolle des Menschen über die Maschine. „Wirklich gut ist ein Pilot, wenn er der Maschine immer ein Stück weit voraus ist, wenn er immer schon mit einkalkuliert, was als nächstes kommen kann."

JEDER SEAT BESITZT EINEN KLAREN CHARAKTER.

Das ist es auch, was er täglich seinem Team von mehr als 1.000 Ingenieuren vermittelt. „Wir arbeiten an der Zukunft, wir entwickeln mehr als fünf Jahre im voraus und definieren dabei begeisternde, erfolgreiche Modelle." Den Weg in diese Zukunft gibt der Leon vor, SEAT wird seine Marke „leonisieren". Das bedeutet, dass alle Zutaten, die den Erfolg dieses Modells ausmachen, in künftigen Produkten zu finden sein werden: Design und Funktionalität, Dynamik und Komfort, Erreichbarkeit und Qualität – all das verbunden mit den besten Technologien, die einen emotionalen Fahrspaß garantieren.

Die volle Konzentration auf den Punkt braucht Matthias Rabe auch auf der Rennstrecke, einer seiner weiteren Leidenschaften. Selbst bei der Abstimmung und den Tests des Rennwagens Leon CUP RACER oder des „zivilen" Topmodells, des Leon CUPRA, war Dr. Rabe aktiv dabei. „SEAT wird als sehr sportliche Marke wahrgenommen", sagt der Technik-Vorstand, „und unsere CUPRA-Palette macht dem alle Ehre."

Die Marke voranbringen, das geht nur mit den bestmöglichen Produkten, weiß Dr. Rabe. Die neuesten Technologien sind hier selbstverständlich, schließlich nutzt SEAT den riesigen Know-how-Pool des Volkswagen-Konzerns. Noch wichtiger aber ist der klare Charakter der Autos. Und der entsteht im perfekten Zusammenspiel von Design und Konzept – und natürlich in diesem ganz speziellen spanisch-deutschen Gen-Pool von SEAT. „In unserer Verbindung aus German Engineering und spanischer Begeisterung sind wir einzigartig", ist Dr. Rabe überzeugt und verweist zum Beispiel auf den Leon ST CUPRA: Der ist nicht nur der Schönste seiner Klasse und auf dem Nürburgring der Schnellste, sondern auf der Straße auch der Sparsamste.

EN
Flying is pure passion for Dr. Matthias Rabe. He loves the absolute control of the machine.

ES
Para Matthias Rabe volar es toda una pasión. Le encanta la sensación de controlar completamente la máquina.

DE
Fliegen ist pure Leidenschaft für Dr. Matthias Rabe. Er liebt die absolute Kontrolle über die Maschine.

SEAT
20V20
THE NEW BRAND SHOWCASE

DE Luca de Meo,
Vorsitzender des Vorstands der SEAT S.A.

DER 20V20 ZEIGT DIE NEUE AMBITION UNSERER MARKE SEAT.

EN THE 20V20 IS A NEW CALLING CARD FOR THE SEAT BRAND.
Luca de Meo, Chairman of the Executive Committee, SEAT S.A.

ES EL 20V20 ES TODA UNA DECLARACIÓN DE INTENCIONES PARA SEAT.
Luca de Meo, presidente del Comité Ejecutivo de SEAT S.A.

EN Dr. Matthias Rabe,
Board Member for Research and Development SEAT S.A.

WE ARE SHOWING WITH THE 20V20 SHOW CAR WHAT A GOOD FIT SUCH A SPORTY SUV IS FOR OUR BRAND.

ES CON EL SHOWCAR 20V20 VAMOS A MOSTRAR
LO BIEN QUE ENCAJA ESTE SUV DEPORTIVO EN NUESTRA MARCA.
Dr. Matthias Rabe, vicepresidente ejecutivo de Investigación y Desarrollo de SEAT S.A.

DE MIT DEM CONCEPT CAR 20V20 ZEIGEN WIR, DASS SOLCH EIN SPORTLICHER
SUV PERFEKT ZU UNSERER MARKE PASST.
Dr. Matthias Rabe, Mitglied des Vorstands für Forschung und Entwicklung SEAT S.A.

ES Alejandro Mesonero-Romanos,
director de Diseño de SEAT

NUESTRO 20V20 TIENE UNA APARIEN-CIA LIMPIA, PERO QUE EVOCA UNA GRAN POTENCIA. TRANSMITE UNA TENSIÓN FENOMENAL Y UNA INTENSA SENSUALIDAD.

EN OUR 20V20 HAS A CLEAR LOOK, WITH POWERFUL FORWARD MOMENTUM.
IT POSSESSES PHENOMENAL TENSION AND GREAT DEPTH OF SENSUALITY.
Alejandro Mesonero-Romanos, Head of SEAT Design

DE UNSER 20V20 HAT EINEN KLAREN AUSDRUCK, EIN STARKES STREBEN
NACH VORNE. ER BESITZT EINE HOHE SPANNUNG UND EINE TIEFE SINNLICHKEIT.
Alejandro Mesonero-Romanos, Leiter SEAT Design

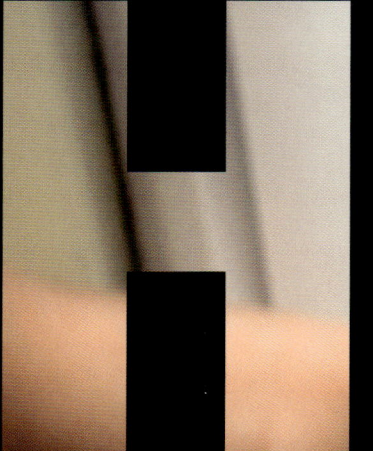

DESIGN PROCESS HOW WE CR

EN 20V20 is a message. It is all about character. It is about the way we at SEAT understand seduction.

ES Con el 20V20 se lanza un mensaje. Es cuestión de carácter. Representa nuestra forma de entender la seducción en SEAT.

DE Der 20V20 hat eine Botschaft. Es geht um Charakter. Und es geht darum, wie wir bei SEAT Verführung verstehen.

SEAT 20V20
DESIGN
PROCESS
EXTERIOR

EN A car is not simply a vehicle; it's not just about transportation, about function. A car is a lot more than that. It's about emotion, seduction, personality, freedom. Good design shows all of that, which is why it needs a clear vision from the very first sketches.

ES Un coche no es simplemente un vehículo, algo funcional que sirve para desplazarse; es mucho más que eso. Se trata de emoción, seducción, personalidad y libertad. Un buen diseño muestra todos estos aspectos; por eso necesita una visión clara desde los primeros bocetos.

DE Ein Auto ist nicht nur ein Fahrzeug. Es geht nicht nur um Transport, um Funktion. Ein Auto ist viel mehr. Es geht um Emotion, um Verführung, um Persönlichkeit, um Freiheit. Gutes Design zeigt all das. Deshalb braucht es eine klare Vision, von der ersten Zeichnung an.

GOOD DESIGN NEEDS A CLEAR
VISION

AND EVERY VISION NEEDS A STRONG
TEAM

EN The 20V20 bears the SEAT design DNA. It is an automobile for passionate, youthful people, created by a team of passionate and creative designers.

ES El 20V20 lleva el ADN del diseño de SEAT. Es un coche para gente apasionada de espíritu joven, creado por un equipo de diseñadores igual de apasionados y creativos.

DE Der 20V20 trägt die Design-DNA von SEAT in die Zukunft. Er ist ein Automobil für passionierte und jung gebliebene Menschen, erdacht von einem Team passionierter und kreativer Designer.

EN Creativity means the implementation of a great many ideas and being open to everything. But then it all has to be condensed and compressed in the interests of clear identity. Thus, the work done on this Concept Car provides enormous impetus to the entire brand.

ES La creatividad consiste en poner en práctica gran cantidad de ideas y estar abierto a todo. Pero después, también hay que condensarlo y sintetizarlo todo para conseguir una identidad clara. El trabajo que se ha realizado en este prototipo supone un impulso enorme para toda la marca.

DE Kreativität bedeutet, eine Vielzahl von Ideen umzusetzen und offen in alle Richtungen zu sein. Dann aber muss wieder verdichtet werden, komprimiert, im Sinne der klaren Identität. So treibt die Arbeit an diesem Concept Car die gesamte Marke kraftvoll voran.

DESIGN
IS CREATIVITY
AND TECHNOLOGY

At SEAT, we believe that practical and functional things can be beautiful – indeed, must be beautiful. That is why, when designing the Concept Car, we also focused on issues like optimum usability and aerodynamics.

En SEAT creemos que las cosas prácticas y funcionales pueden y deben ser bonitas. Por eso, cuando diseñamos el prototipo también nos fijamos en aspectos como la óptima facilidad de uso y la aerodinámica.

Wir bei SEAT sind überzeugt, dass auch praktische und funktionale Dinge schön sein können. Ja, schön sein müssen. Deshalb achten wir auch beim Design des Concept Cars auf Themen wie optimalen Nutzwert oder Aerodynamik.

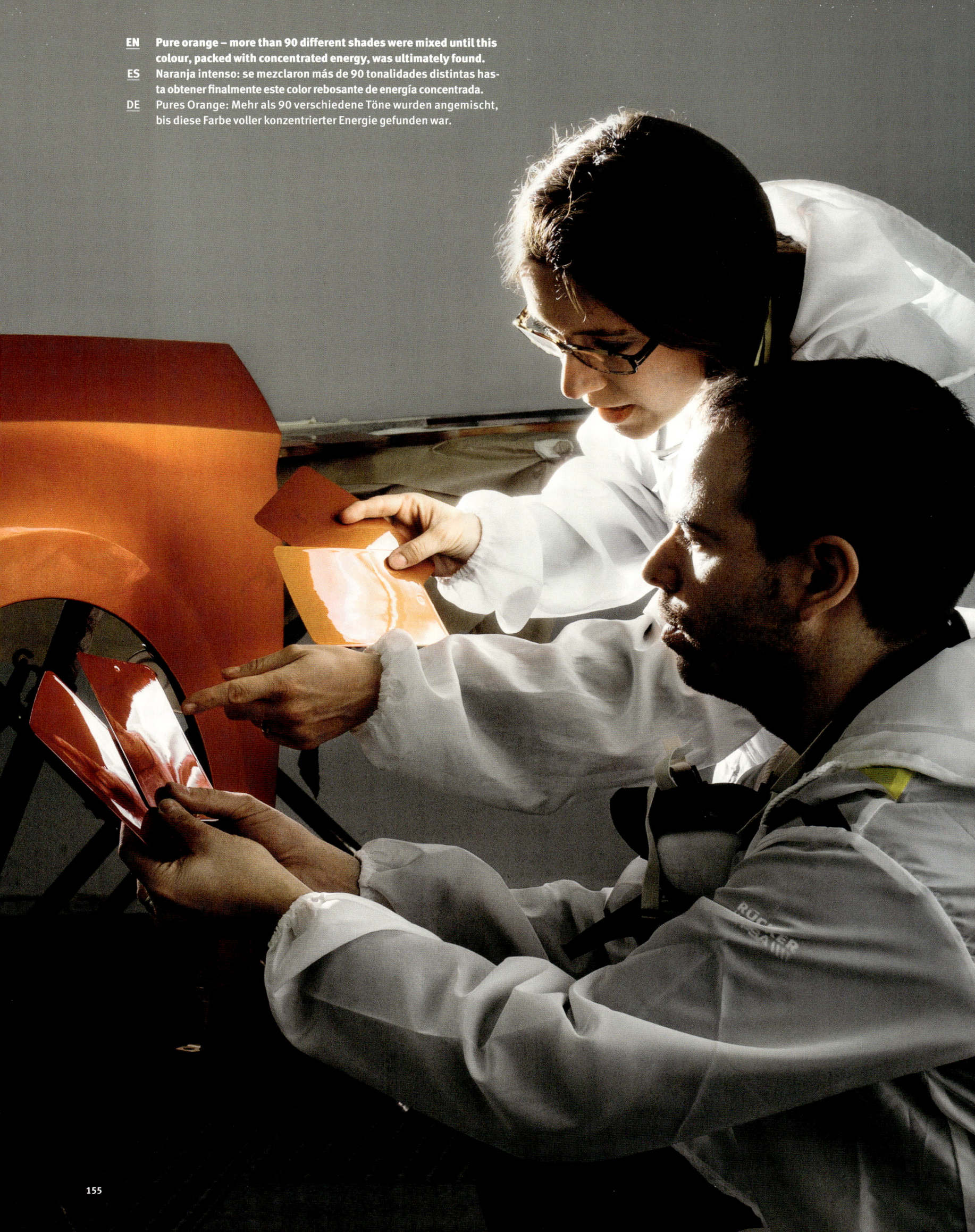

ROUGH & SMOOTH

EN — In the 20V20, sports utility means the combination of fire and silk, of robustness and elegance. It is ready for whatever lies ahead, but always retains its style.

ES — En el 20V20, el concepto de deportivo utilitario aúna fuego y cuero, solidez y elegancia. Está listo para enfrentarse a cualquier cosa sin perder su estilo.

DE — Sports Utility bedeutet beim 20V20 die Kombination aus Feuer und Seide, aus Robustheit und Eleganz. Er ist bereit für alle Wege, aber bewahrt stets seinen Stil.

EN **Mission accomplished: Countless days and nights of intensive
work have flowed into this one Concept Car. Now it is perfect – and
ready for its grand entrance onto the motor show stage.**

ES Misión cumplida: este prototipo único es el resultado de un sinfín de
días y noches de intenso trabajo. Ahora que es perfecto ya está
listo para entrar por la puerta grande en salones y ferias del automóvil.

DE Mission accomplished: Ungezählte Tage und Nächte intensiver
Arbeit stecken in diesem einen Concept Car. Nun ist es perfekt – und
bereit für den ganz großen Auftritt auf den Bühnen der Automobilmessen.

EX
TER
IOR

PASSIONATE
BEAUTY

SEAT
20V20

With the 20V20, we have turned a dream into reality. We did this with our SEAT design DNA, with our love of beautiful things, attention to detail and to perfection.

DE Mit dem 20V20 haben wir einen Traum in Realität verwandelt – mit unserer SEAT Design-DNA. Mit unserer Liebe zu schönen Dingen, zum Detail und zur Perfektion.

ES El 20V20 es un sueño hecho realidad. Lo hemos conseguido gracias al ADN del diseño de SEAT, a nuestra pasión por la belleza y a nuestra atención a los detalles y la perfección.

EN The 20V20 stands high on the road, but its body remains low and muscular. This is our idea of a coupé on big wheels. The front displays the sign of the X; all lines seem to converge on the large SEAT logo. What this shows is that we are proud of our logo.

ES El 20V20 destaca en la carretera, pero su carrocería es baja y con aspecto musculoso. Es nuestra idea de cómo debe ser un cupé sobre grandes ruedas. En la parte frontal, las líneas visuales forman una X que parece converger en un gran logotipo de SEAT. Es un símbolo de nuestro orgullo por la marca.

DE Der 20V20 steht hoch über der Straße, aber sein Körper bleibt flach und muskulös. Das ist unsere Idee von einem Coupé auf großen Rädern. Die Front zeigt das Motiv des X, alle Linien scheinen in dem großen SEAT Logo zusammenzulaufen. Das zeigt: Wir sind stolz auf unser Logo.

PERFECTION IN EVERY DETAIL

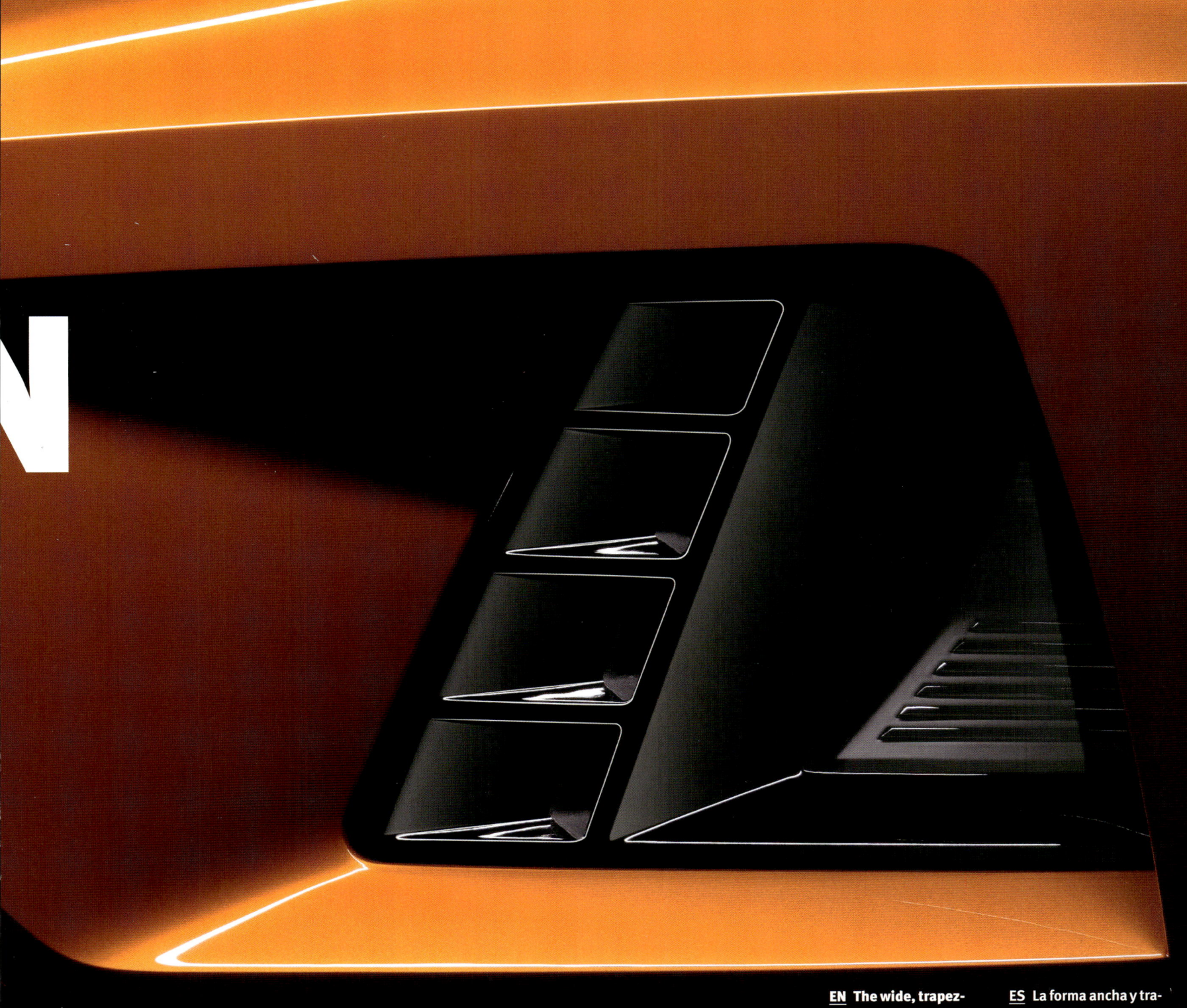

EN The wide, trapezoidal form of the front bumper conveys stability and power. Every one of its details has been sculpturally crafted, instilling the beholder with a compelling sense of precisely balanced forms and colours, surfaces and materials.

ES La forma ancha y trapezoidal del parachoques delantero transmite estabilidad y potencia. Cada uno de los detalles se ha confeccionado de forma escultórica y el conjunto aporta una tentadora sensación de formas y colores, superficies y materiales equilibrados de forma precisa.

DE Mit seiner breiten Trapezform vermittelt der vordere Stoßfänger Stabilität und Kraft. Dabei ist er in jedem Detail skulpturhaft ausgearbeitet und begeistert durch präzise aufeinander abgestimmte Formen und Farben, Oberflächen und Materialien.

AT SEAT WE KNOW THAT FUNCTIONAL CAN BE BEAUTIFUL

EN The door handle shows the SEAT design DNA at its best – minimalist, sculptural in form, precise lines, harmony in light and shade – and perfectly executed.

ES Los tiradores de las puertas son una muestra del ADN del diseño de SEAT en todo su esplendor: minimalistas, de formas escultóricas y líneas precisas, con armonía de luces y sombras y ejecutados a la perfección.

DE Der Türgriff zeigt die SEAT Design-DNA at its best: Reduzierte, skulpturhafte Form, präzise Linien, harmonisch in Licht und Schatten – und perfekt umgesetzt.

HIGH-TECH BEAUTY

EN A SEAT is dynamic and enormous fun to drive. Its design reflects this sporty spirit and shows the technology involved, such as the high-performance brake system behind the fine spokes of the large wheels.

ES Un SEAT es dinámico y divertido de conducir. Su diseño refleja este espíritu deportivo y muestra abiertamente su tecnología, como el sistema de frenado de alto rendimiento que dejan entrever los radios de las grandes ruedas.

DE Ein SEAT ist dynamisch und bringt größten Fahrspaß. Sein Design reflektiert diesen sportlichen Geist und zeigt die Technologie dazu, etwa die leistungsstarke Bremsanlage unter den feinen Speichen der großen Räder.

SEAT MEANS INTELLIGENCE WITH CHARACTER

EN A SEAT has character – the character of an individual who knows exactly what he wants – which makes it the perfect match for the personality of its driver.

DE Ein SEAT besitzt Charakter. Den Charakter eines Menschen, der genau weiß, was er will. Und damit passt er perfekt zur Persönlichkeit seines Fahrers.

ES Los coches SEAT tienen el carácter de una persona que sabe perfectamente lo que quiere; esto encaja a la perfección con la personalidad de quien lo conduce.

EN The character lines and the dual blisters give the side its wonderful tension. The low roofline, the narrow glass surfaces and the muscular body of the well-trained athlete. The rear lights continue this tension – like arrows about to be fired.

ES Las líneas de carácter y los pliegues dobles de la carrocería confieren a los laterales una magnífica tensión. A esto se suma el techo bajo, los finos cristales y el cuerpo musculoso de un atleta bien entrenado. Los faros traseros, en forma de flechas a punto de salir disparadas, enfatizan la tensión.

DE Die Charakterlinien, die doppelten Blister, geben der Seite ihre wunderbare Spannung. Das flache Dach, die schmalen Glasflächen, dazu der muskulöse Körper des trainierten Athleten. Die Heckleuchten nehmen diese Spannung auf – wie Pfeile kurz vor dem Abschuss.

ALEJANDRO MESONERO-ROMANOS' PERSPECTIVE

WE LOVE CARS AT SEAT.
YOU CAN SEE IT AND YOU CAN FEEL IT.

ES En SEAT nos encantan los coches, es algo que se ve y se siente.
DE Wir bei SEAT lieben Autos. Das sieht man ihnen an, und das spürt man.

EN
The place by the sea stands
for expansiveness of
thought and inspiration
from simplicity.

ES
Un lugar junto al mar facilita
abrir la mente e
inspirarse en las cosas
sencillas.

DE
Der Platz am Meer steht für
die Weite der Gedanken
und die Inspiration aus der
Einfachheit.

EN Alejandro Mesonero-
    Romanos Aguilar,
    Head of SEAT Design
ES Alejandro Mesonero-
    Romanos Aguilar,
    Director de Diseño SEAT S.A.
DE Alejandro Mesonero-
    Romanos Aguilar,
    Leiter SEAT Design

EN It is the expanse, the emptiness of space that Alejandro Mesonero-Romanos likes so much about his personal place of inspiration. "Our world is often so full, fast and loud. That's when inspiration also comes from simplicity, from slowness." Sitting on the beach, feeling the gentle breeze and breathing in the smell of the water, watching the distinct vibrancy of the sea – and enjoying a coffee. Consciously experiencing the simple things in life – and enjoying them to the full.

Clear and simple, but aware and intensive – for the head of SEAT Design, this describes the Mediterranean lifestyle, which also has a major influence on the car brand from Barcelona. The sheer love of life, the appreciation of pleasures great and small is part of that, as is a distinct feel for style and quality. "After all, cars are more than just means of transport. They are something emotionally highly complex."

SEAT MEANS PRECISION AND SENSUALITY IN EQUAL MEASURE.

"And we should enjoy them," is the firm belief of Alejandro Mesonero-Romanos. "Cars are about form and power; cars are about living and experiencing. They give their owners a certain attitude to life and express an aspect of their personality."

SEAT lives and breathes these values – something that is completely clear to the design boss. Every SEAT is functionally a very good car, with the latest technology and the very best craftsmanship. "But equally important is what its owner sees, what he feels, how the design, the materials and the way they are crafted bring pleasure every single day." It is not about the big show for the moment, but about enduring quality, the best form of sustainability. "And, in terms of design, this often lies in simplicity and clarity. You don't always have to fill surfaces and space with lines. Lines should always be applied with specific intent, carefully dosed like good seasoning."

Alejandro Mesonero-Romanos values not only the hours spent at the sea, he also loves race tracks. Although here at the Autódromo de Terramar, an oval circuit built 90 years ago, there has been no active driving for many years, the smell of petrol and rubber still seems to hang in the air. "I love cars. We at SEAT love cars. You can see it and you can feel it."

ES Lo que a Alejandro Mesonero-Romanos le gusta más de su lugar de inspiración es la amplitud y el vacío del espacio. «Nuestro mundo suele estar lleno de cosas, es ruidoso y rápido. Por eso la inspiración surge también de la sencillez, de la lentitud». Sentarse en la playa, sentir la brisa e inhalar una bocanada de aire marino, observar el brillo nítido del agua y disfrutar de un café. Experimentar las cosas sencillas de la vida y disfrutarlas al máximo.

Es algo claro y sencillo, pero también consciente e intenso. Para el jefe de Diseño de SEAT, esto describe el estilo de vida mediterráneo que tanto influye en la marca de coches de Barcelona. El auténtico amor por la vida, el gusto por los pequeños y los grandes placeres forman parte de esta forma de vivir, que se traduce en una manera personal de percibir el estilo y la calidad.

SEAT ES PRECISIÓN Y SENSUALIDAD A PARTES IGUALES.

«Al fin y al cabo, los coches son más que simples medios de transporte. Son muy complejos desde el punto de vista emocional y deberíamos disfrutar de ellos». Es algo que Alejandro cree con firmeza. «Los coches tienen que ver con la forma y la potencia, con vivir y experimentar. Hacen que sus dueños afronten la vida de un modo concreto y son un reflejo de su personalidad».

El jefe de Diseño tiene muy claro que en SEAT se viven y se respiran estos valores. Todos los SEAT son muy buenos coches desde el punto de vista funcional y cuentan con la última tecnología y la mejor producción artesanal. «Pero igual de importante es lo que su dueño ve, lo que siente, el placer que le proporcionan cada día el diseño, los materiales y el modo en que están confeccionados». No se trata de hacer una breve aparición estelar, sino de mantener la calidad, que es el mejor tipo de sostenibilidad. «Y por lo que se refiere al diseño, esto suele surgir de la sencillez y la claridad. No hace falta llenar siempre el espacio de líneas. Las líneas deberían utilizarse con un fin concreto y dosificarse con cuidado, como un buen condimento».

Alejandro Mesonero-Romanos aprecia las horas que pasa junto al mar y le encantan las pistas de carreras. En el circuito oval del Autódromo de Terramar, construido hace 90 años, no hay actividad desde hace mucho tiempo, pero el olor a gasolina y neumático todavía parece flotar en el aire. «Me encantan los coches. En SEAT nos encantan los coches, es algo que se ve y se siente».

DE Es ist die Weite, die Leere des Raums, die Alejandro Mesonero-Romanos an seinem ganz persönlichen Ort der Inspiration so gefällt. „Unsere Welt ist oft so voll, schnell und laut. Dann liegt die Inspiration auch in der Einfachheit, in der Langsamkeit." Am Strand sitzen, die angenehme Brise spüren und den Geruch des Wassers einatmen, die ganz eigene Lebendigkeit des Meeres beobachten – dazu einen Kaffee trinken. Die einfachen Dinge bewusst erleben – und intensiv genießen.

Klar und einfach, aber bewusst und intensiv. Das beschreibt für den Leiter von SEAT Design den mediterranen Lebensstil, der auch die Automobilmarke aus Barcelona deutlich prägt. Der Spaß am Leben, der Genuss der kleinen und großen Freuden gehören dazu ebenso wie ein eindeutiges Gefühl für Stil und Qualität. „Autos sind doch nicht nur Transportmittel. Sie sind etwas emotional sehr Komplexes."

SEAT BEDEUTET PRÄZISION UND SINNLICH- KEIT GLEICHERMASSEN.

„Und man sollte sie genießen", ist Alejandro Mesonero-Romanos überzeugt. „Autos sind Form und Kraft, Autos sind Leben und Erleben. Sie schenken ihrem Besitzer ein Lebensgefühl und sie drücken einen Teil seiner Persönlichkeit aus."

SEAT lebt diese Werte, das ist für den Designchef völlig klar. Jeder SEAT ist ein funktional sehr gutes Automobil, mit modernster Technik und bester Verarbeitung. „Aber ebenso entscheidend ist, was sein Besitzer sieht, was er spürt, wie das Design, die Materialien und ihre Verarbeitung jeden Tag von neuem Freude machen." Es geht nicht um die große Show für den Moment, sondern um dauerhafte Qualität, um die beste Form von Nachhaltigkeit. „Und die liegt auch im Design oft in der Einfachheit und Klarheit. Man muss Flächen und Räume nicht immer mit Linien füllen. Man sollte Linien stets bewusst einsetzen, dosiert wie gute Gewürze."

Alejandro Mesonero-Romanos schätzt nicht nur die Stunden am Meer, er ist auch gerne auf Rennstrecken. Zwar wird hier im Autódromo de Terramar, einem vor 90 Jahren entstandenen Ovalkurs südlich von Barcelona, seit vielen Jahren nicht mehr aktiv gefahren, der Geruch von Benzin und Gummi scheint aber immer noch in der Luft zu liegen. „Ich liebe Autos. Wir bei SEAT lieben Autos. Das sieht man ihnen an, und das spürt man."

EN
The Autódromo de Terramar close to Barcelona has long ceased to be an active racing venue. But the smell of petrol and the heat of the engines still seem to hang in the air. The head of SEAT Design loves this place of inspiration.

ES
Hace mucho tiempo que el Autódromo de Terramar ya no es un circuito de carreras activo, pero el olor a gasolina y el calor de los motores todavía parecen flotar en aire. Al jefe de Diseño de SEAT le parece un lugar tremendamente inspirador.

DE
Für aktives Racing hat das Autódromo de Terramar in der Nähe von Barcelona längst ausgedient. Aber der Geruch von Benzin und die Hitze der Motoren scheinen noch immer in der Luft zu liegen. Der SEAT Designchef liebt diesen Ort der Inspiration.

THE CONSCIOUS APPRECIATION OF GOOD THINGS — THAT IS THE MEDITERRANEAN WAY OF LIFE.

ES El estilo de vida mediterráneo se basa en la valoración consciente de las cosas buenas de la vida.
DE Gute Dinge bewusst genießen. Das ist mediterrane Lebensart.

EN The interior design of the 20V20 is a harmonious blend of expressiveness, functionality and precision.

ES El diseño interior del 20V20 es una combinación armoniosa de expresividad, funcionalidad y precisión.

DE Das Interieurdesign des 20V20 ist eine harmonische Kombination aus Ausdrucksstärke, Funktionalität und Präzision.

HOW A CRE LIVING SPA

SEAT 20V20 DESIGN PROCESS INTERIOR

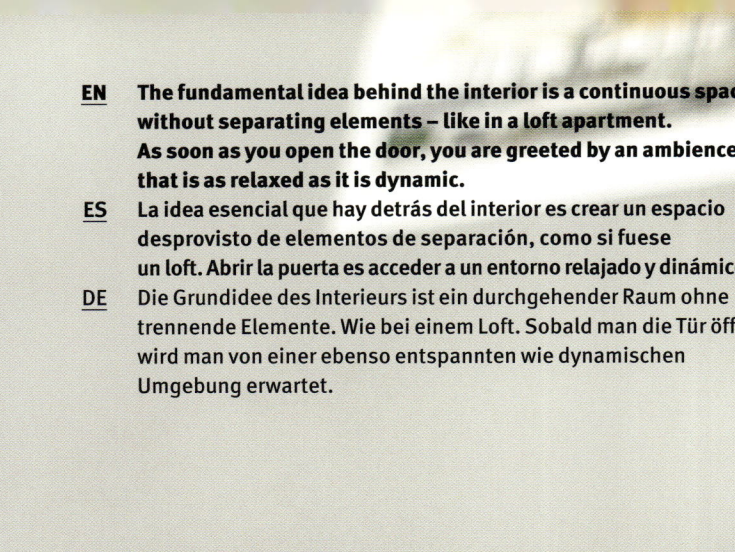

EN The fundamental idea behind the interior is a continuous space without separating elements – like in a loft apartment. As soon as you open the door, you are greeted by an ambience that is as relaxed as it is dynamic.

ES La idea esencial que hay detrás del interior es crear un espacio desprovisto de elementos de separación, como si fuese un loft. Abrir la puerta es acceder a un entorno relajado y dinámico.

DE Die Grundidee des Interieurs ist ein durchgehender Raum ohne trennende Elemente. Wie bei einem Loft. Sobald man die Tür öffnet, wird man von einer ebenso entspannten wie dynamischen Umgebung erwartet.

A WELCOMING SPACE –
FROM THE FIRST
SKETCH

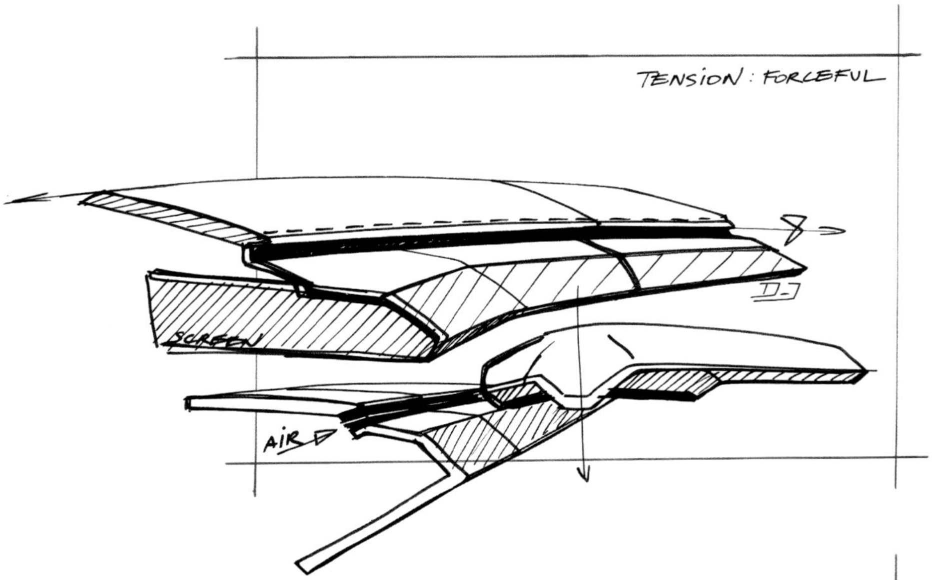

TENSION : FORCEFUL

SCREEN

AIR

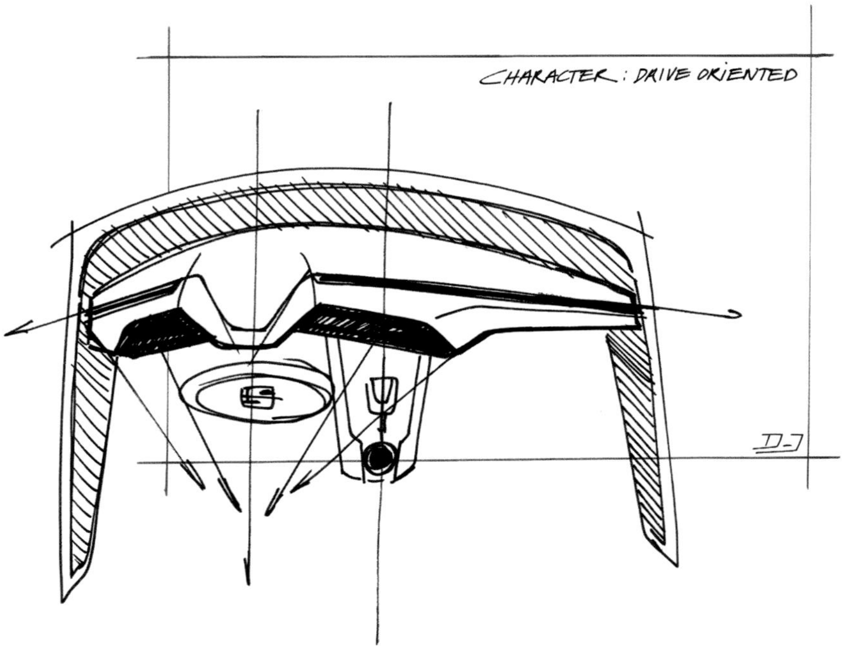

CHARACTER : DRIVE ORIENTED

DRIVER ORIENTED
FOR MAXIMUM PERFORMANCE

FLOATING DASHBORD / CONSOLE

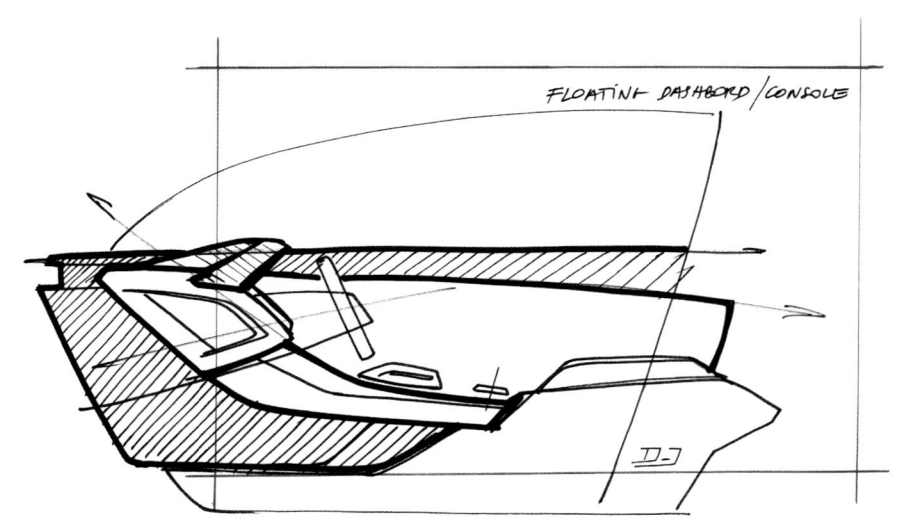

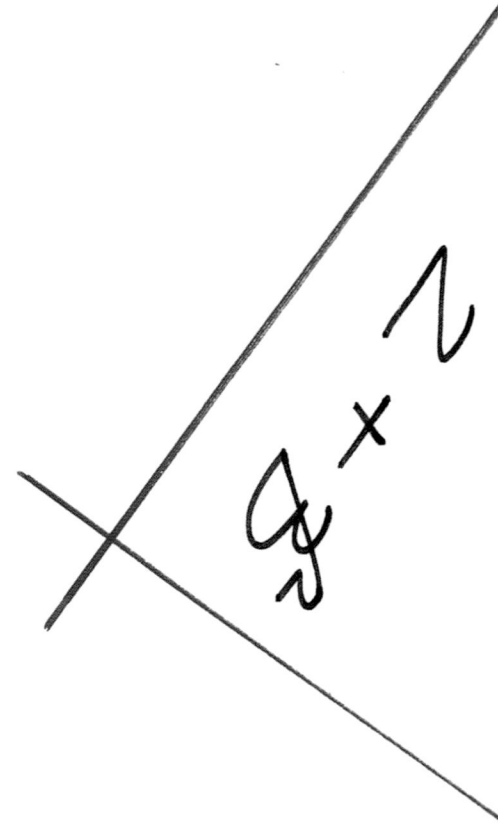

EN The driver is the focal point. He needs to be in full control of the vehicle at all times. The dashboard and centre console are clearly oriented towards him, the operating elements are closely grouped, with all displays only a glance away.

ES El centro de atención es el conductor, que debe tener el control absoluto del vehículo en todo momento. El cuadro de instrumentos y la consola central están claramente orientados hacia él. Los elementos operativos están agrupados en el mismo espacio y basta con levantar la vista para ver todas las pantallas.

DE Der Fahrer steht im Mittelpunkt, er braucht jederzeit die volle Kontrolle über das Fahrzeug. Instrumententafel und Mittelkonsole sind klar auf ihn orientiert, die Bedienelemente liegen nahe beisammen, alle Anzeigen sind nur eine Augenbewegung entfernt.

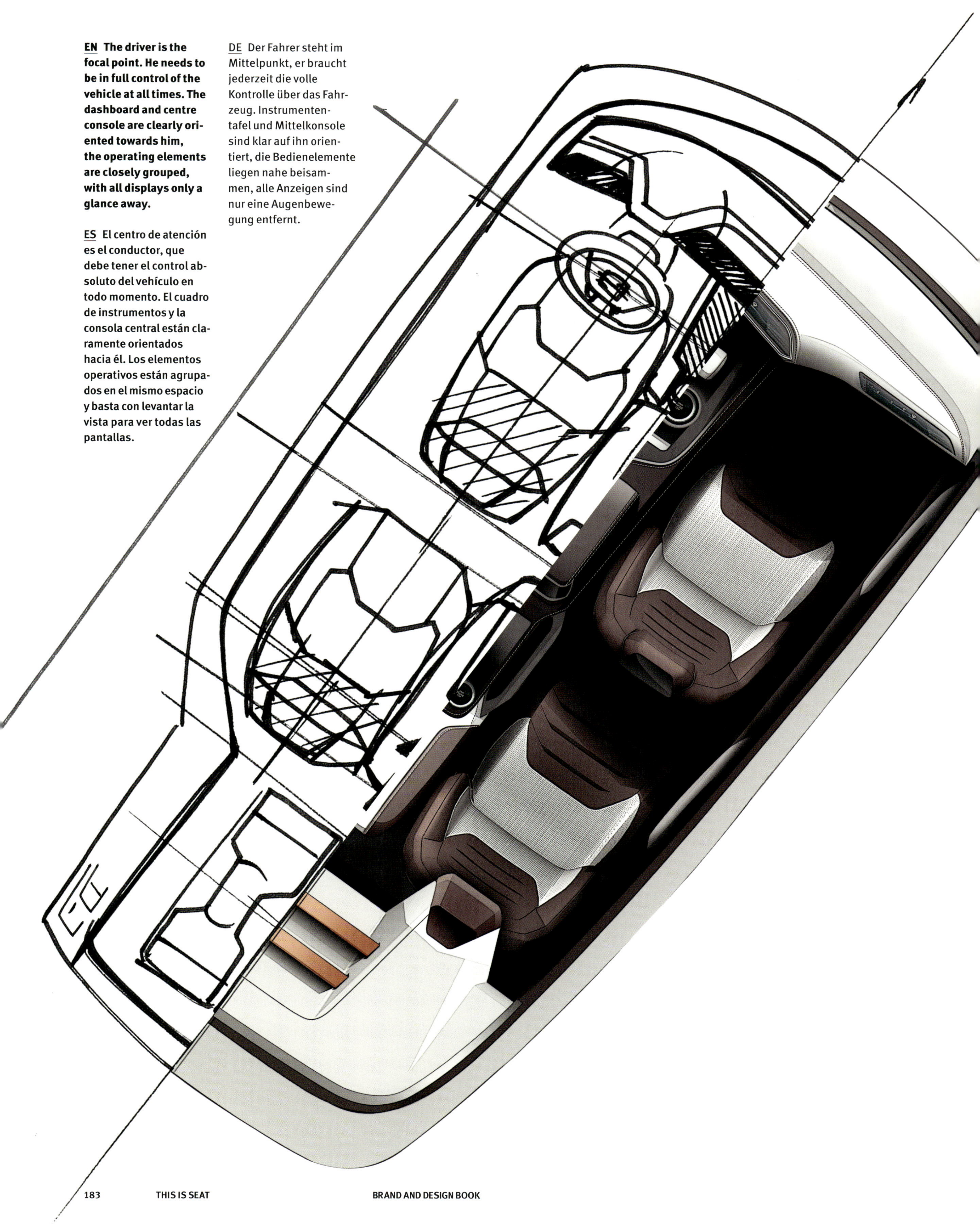

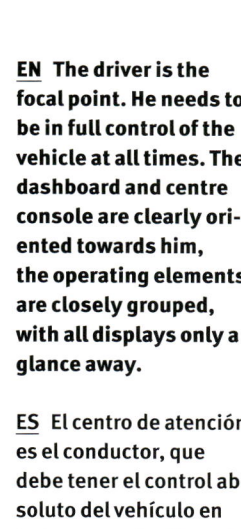

THE OPTIMUM IS ALWAYS ONE STEP AWAY

EN It takes the three-dimensional model to show how well an idea really works. And then every single millimetre is scrutinised – until the optimum is reached.

ES Para ver si una idea funciona es necesario contar con un modelo en tres dimensiones. Después, se examina cada milímetro hasta estar seguros de que el resultado es insuperable.

DE Erst das dreidimensionale Modell zeigt, wie gut die Idee wirklich funktioniert. Und dann wird um jeden Millimeter gerungen – bis das Optimum erreicht ist.

EN Is it still a skill or is it art? Turning the sketches into a model is possible only with extremely deft hands.

ES ¿Técnica o arte? Convertir los bocetos en un modelo requiere una gran destreza manual.

DE Noch Handwerk oder schon Kunst? Die Umsetzung von der Skizze zum Modell gelingt nur mit äußerst geschickten Händen.

CLAY ARTISTS

EN It takes a perfect feel for style – the choice of colours and materials calls for enormous sensitivity and a finely tuned instinct for harmony.

ES Es necesario tener un sentido del estilo depurado: la elección de los colores y los materiales exige una gran sensibilidad y sentido de la armonía.

DE Eine Frage perfekten Stilgefühls: Die Auswahl der Farben und Materialien braucht eine hohe Sensibilität und viel Gespür für Harmonie.

COLOUR AND TRIM

EN Fabric, leather, glass, metal – the materials determine what is seen and felt. Their surfaces form the interface for human contact – where quality can be touched, e.g. by running a hand along the warm, soft, natural leather. The colours reflect the Mediterranean atmosphere around Barcelona.

ES Tejidos, cuero, vidrio y metal: los materiales determinan lo que se percibe con la vista y el tacto. Las superficies son el punto de contacto entre el objeto y la persona y permiten palpar la calidad. Por ejemplo, pasando la mano por el cuero natural, cálido y suave. Los colores son un reflejo del ambiente mediterráneo de Barcelona.

DE Stoff, Leder, Glas, Metall. Die Materialien bestimmen, was man sieht und spürt. Ihre Oberflächen schaffen den Kontakt zum Menschen. Und der fühlt den Wert, etwa beim Berühren des warmen und weichen, naturbelassenen Leders. Die Farben spiegeln die mediterrane Atmosphäre rund um Barcelona.

ROUGH & SMOOTH

BRAND AND DESIGN BOOK

PASSION

FEEL
THE SHAPE

HOW WE CREATE
CONNECTIVITY

EN The 20V20 communicates with its driver via the instruments. The designers have created a virtual cockpit – with individual surfaces that can easily be adapted to suit personal preferences.

ES El 20V20 se comunica con el conductor a través de los instrumentos. Los diseñadores han creado una cabina de mandos virtual cuyas superficies pueden adaptarse fácilmente a las preferencias de cada conductor.

DE Über die Instrumente kommuniziert der 20V20 mit seinem Fahrer. Die Designer haben ein Virtual Cockpit geschaffen – mit individuellen Oberflächen, leicht adaptierbar an persönliche Vorlieben.

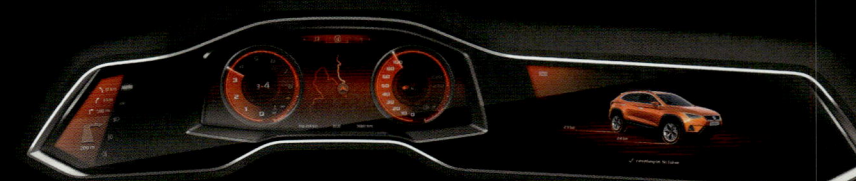

THE HEART OF 20V20

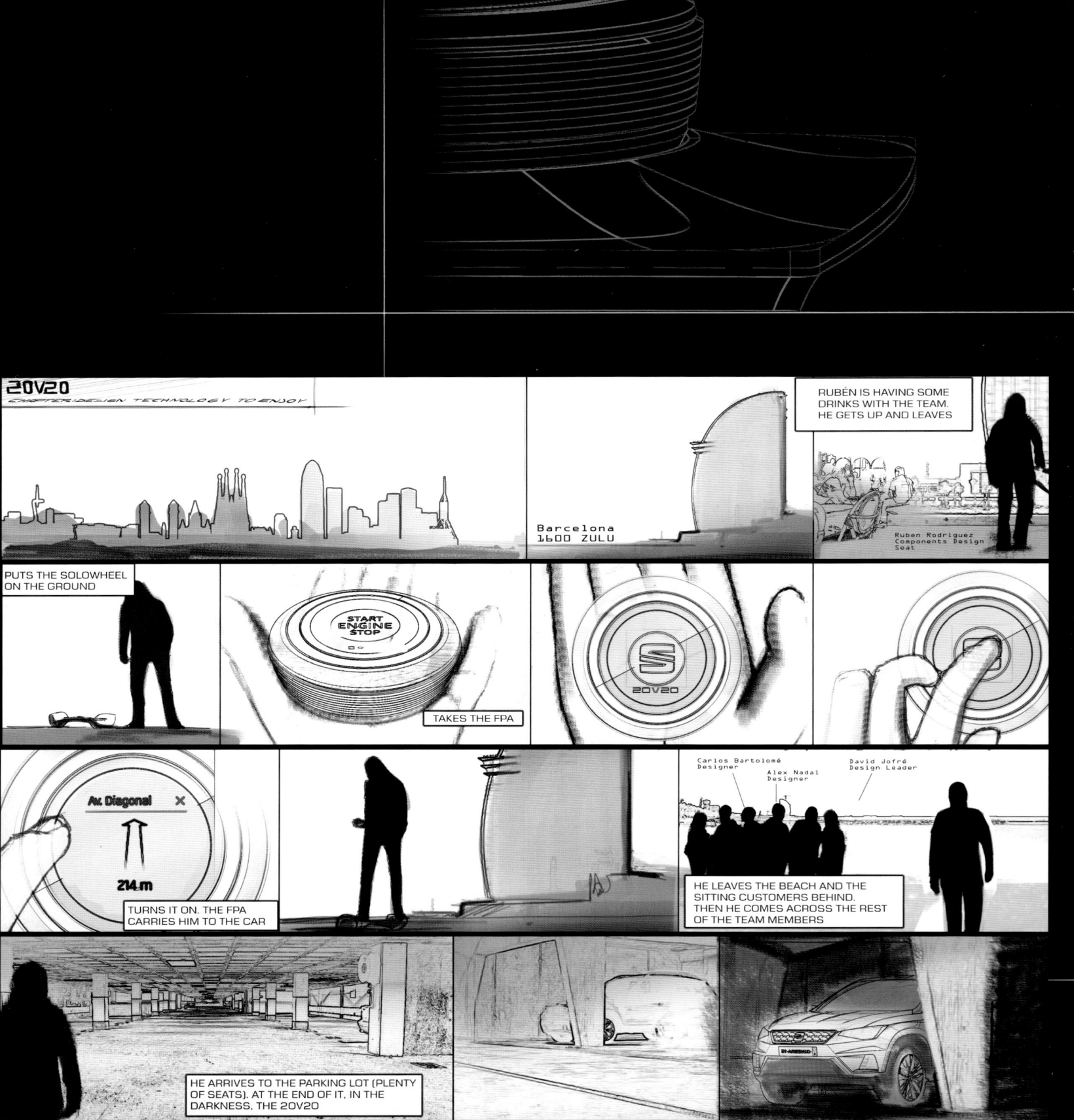

EN The idea for the "heart" of the 20V20 – an individual, mobile operating element for a diverse array of functions.

ES La idea del «corazón» del 20V20: un elemento de control individual y móvil que concentra todo un abanico de funciones.

DE Die Idee für das „Herz" des 20V20: ein individuelles, mobiles Bedienelement für vielfältige Funktionen.

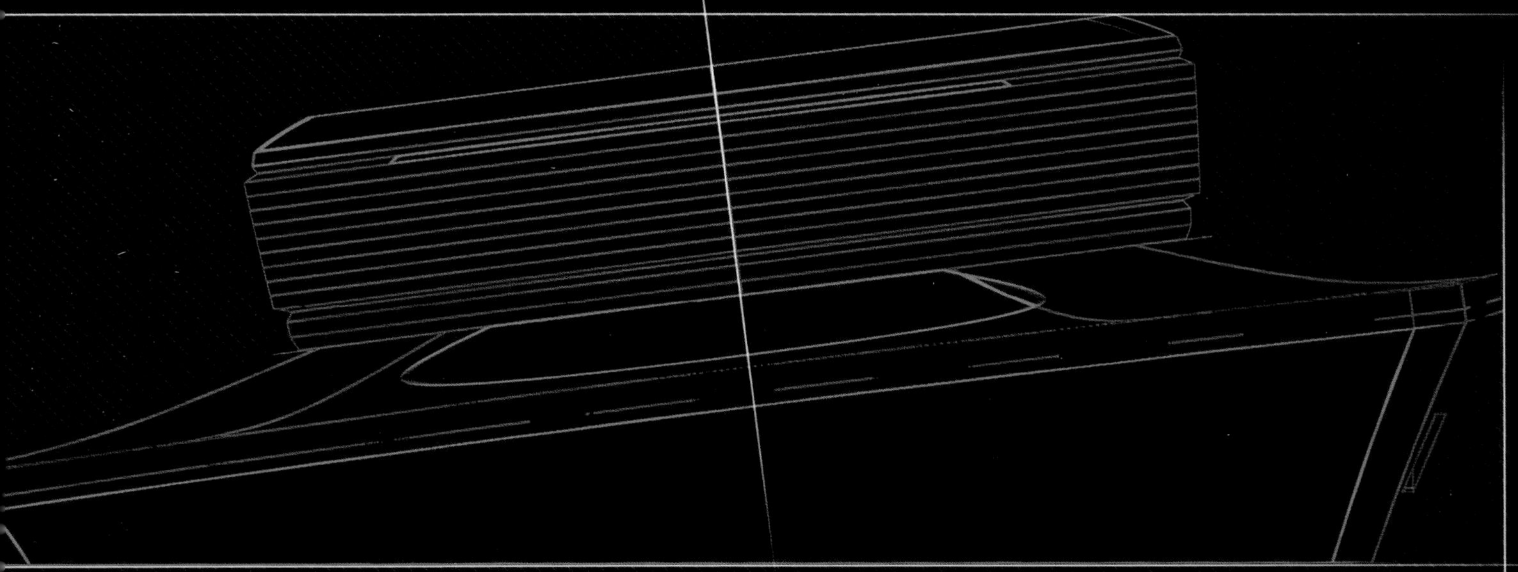

THE WELCOME CEREMONY STARTS

DETAIL OF THE REARVIEW MIRROR AND PROJECTION OF THE WELCOME MESSAGE ON THE GROUND

THE TRUNK OPENS AUTOMATICALLY AND RUBÉN PLACES THE SOLOWHEEL INSIDE

IMAGE OF THE HMI DISPLAYING "DRINK WARMED UP TO 26°", FOLLOWED BY AND IMAGE OF THE BOTTLE AREA. THE AMBIENCE LIGHT FOCUSES ON THIS SPOT

HE PLACES THE FPA

VIDEO OF THE FPA CONCEPT (FLYING PIECES, ETC.)

DOOR AMBIENCE LIGHT

THEN HE PRESSES THE FPA AND THE CAR TURNS ON

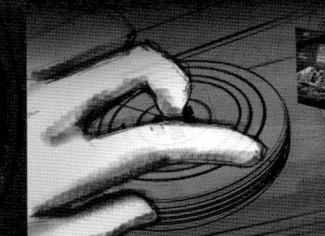
START ENGINE STOP

HE SWITCHES BETWEEN DIFFERENT MODES

USER
INTERFACE

EN The proverbial key to driving pleasure in the 20V20 is the SEAT Personal Drive Device, a beautiful piece of tactile design in the shape of a large coin. Outside the vehicle, this element functions as a mobile navigation system, as well as a remote control for functions such as stationary heating or cooling and for managing the charging process in the case of the plug-in hybrid version. When in the vehicle, the SEAT Personal Drive Device is mounted on the centre console using magnets and serves as an operating element for the SEAT Drive Profile individualisation functions.

ES La clave para el placer de la conducción del 20V20 es el SEAT Personal Drive, un bonito dispositivo táctil en forma de moneda grande. Fuera del vehículo, este elemento funciona como un sistema de navegación móvil, y también como control remoto para calentar o refrigerar el vehículo mientras está aparcado o para gestionar el proceso de carga en la versión híbrida enchufable. Dentro del vehículo, el dispositivo se monta en la consola central mediante imanes, y sirve como elemento de control para las funciones de personalización del SEAT Drive Profile.

DE Den sprichwörtlichen Schlüssel zum Fahrvergnügen im 20V20 liefert das SEAT Personal Drive Device, ein fein gestalteter „Handschmeichler" in der Form einer großen Münze. Außerhalb des Fahrzeugs funktioniert dieses Element als mobiles Navigationssystem, aber auch als Fernbedienung etwa für Standheizung und -kühlung oder den Ladevorgang im Falle der Plug-in-Hybridversion. Im Fahrzeug wird das Personal Device von Magneten auf der Mittelkonsole gehalten – als Bedienelement für die Individualisierungsfunktionen des SEAT Drive Profile.

EN Individual screen designs and the warm ambient lighting create a feeling of wellbeing in the 20V20.

ES Las pantallas individuales y la cálida iluminación ambiente del 20V20 producen una agradable sensación de bienestar.

DE Individuelle Screendesigns und die warme Ambientebeleuchtung schaffen Wohlfühlstimmung im 20V20.

IN TER IOR

PURE QUALITY AND REFINEMENT

EN Premium materials are used in innovative ways and beautifully crafted to create an inviting interior. One example is the saddle brown natural leather by Italian brand Poltrona Frau.

ES Los materiales de alta calidad se utilizan de formas innovadoras y se trabajan con precisión para crear un interior acogedor. Un ejemplo es el cuero natural de color marrón para los asientos de la firma italiana Poltrona Frau.

DE Hochwertige Materialien werden innovativ eingesetzt und bestens verarbeitet. Sie schaffen ein einladendes Interieur. Ein Beispiel ist das sattelbraune Naturleder der italienischen Marke Poltrona Frau.

FEEL THE QUALITY

EN Metal, leather, glass, high-gloss plastic – a SEAT is there to be touched, a pleasure, day after day.

ES Metal, cuero, vidrio y plástico de brillo intenso. Tocar un SEAT es un placer que se renueva cada día.

DE Metall, Leder, Glas, hochglänzender Kunststoff – einen SEAT zu berühren, ist ein Vergnügen. Jeden Tag aufs Neue.

EN The SEAT DNA is unmistakeable – the spacious ambience, the horizontal cockpit, the large touchscreen. And yet everything has been systematically further developed and is on a whole new level. SEAT is a brand in motion.

ES El ADN de SEAT es inconfundible: un interior espacioso con una cabina horizontal y una gran pantalla. Cada elemento se ha desarrollado progresivamente hasta alcanzar un nivel completamente nuevo. SEAT es una marca en continuo movimiento.

DE Die SEAT DNA ist unverkennbar. Das geräumige Ambiente, das horizontale Cockpit, der großzügige Touchscreen. Und doch ist alles auf einer neuen Stufe und konsequent weiterentwickelt. SEAT ist eine Marke in Bewegung.

THE LEATHER, THE COLOURS – ALL OF IT RADIATES MEDI- TERRANEAN SPIRIT.

EN Colours and materials are inspired by the everyday surroundings of the designers – from the Mediterranean atmosphere around Barcelona. SEAT 20V20 – a car that could only have been created here.

ES Los colores y los materiales están inspirados en el entorno de los diseñadores: el ambiente mediterráneo de Barcelona y sus alrededores. El SEAT 20V20 es un coche que solo podía nacer en un lugar así.

DE Farben und Materialien sind inspiriert von der alltäglichen Umgebung der Designer, von der mediterranen Atmosphäre rund um Barcelona. SEAT 20V20 – ein Automobil, das nur hier entstehen konnte.

EN The "heart" of the 20V20 is not only a multi-functional, mobile operating unit. It is, above all, a haptic experience, a perfectly crafted and wonderfully tactile object. You have a precious piece of the 20V20 with you at all times – even when the car is parked in the garage.

ES El «corazón» del 20V20 no es solo una unidad de control móvil multifuncional. Este dispositivo es, ante todo, una experiencia háptica, concentrada en un objeto perfecto y extremadamente táctil. Una valiosa pieza del 20V20 que se puede utilizar en cualquier momento, incluso cuando el coche está aparcado.

DE Das „Herz" des 20V20 ist nicht nur ein multifunktionales, mobiles Bedienteil. Es ist vor allem ein haptisches Erlebnis, ein perfekt verarbeiteter „Handschmeichler". So hat man immer ein wertvolles Stück vom 20V20 bei sich – auch wenn das Auto gerade mal in der Garage parkt.

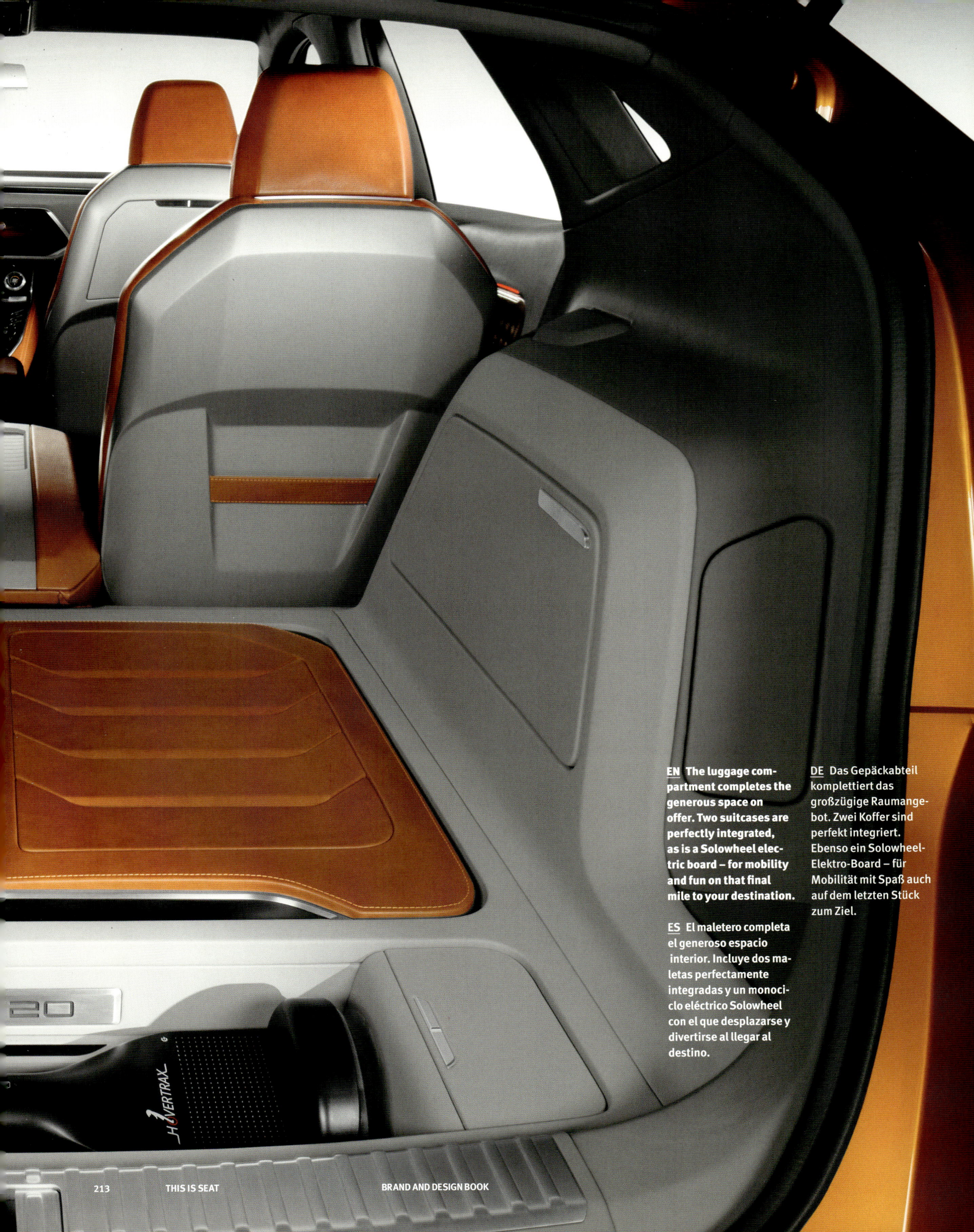

EN The luggage compartment completes the generous space on offer. Two suitcases are perfectly integrated, as is a Solowheel electric board – for mobility and fun on that final mile to your destination.

ES El maletero completa el generoso espacio interior. Incluye dos maletas perfectamente integradas y un monociclo eléctrico Solowheel con el que desplazarse y divertirse al llegar al destino.

DE Das Gepäckabteil komplettiert das großzügige Raumangebot. Zwei Koffer sind perfekt integriert. Ebenso ein Solowheel-Elektro-Board – für Mobilität mit Spaß auch auf dem letzten Stück zum Ziel.

CONNECTED TO LIFE

EN Be connected – with the world and with your friends, all the time and everywhere. This has become taken for granted as part of the young, urban lifestyle – just like the automobile itself. SEAT brings both together with convenient and secure connectivity solutions.

ES Siempre conectados con el mundo y con nuestros amigos, todo el tiempo y en cualquier lugar. Es algo característico de un estilo de vida joven y urbano, como este coche. SEAT acerca ambos mundos mediante soluciones de conectividad cómodas y seguras.

DE Be connected – immer und überall mit der Welt und seinen Freunden verbunden sein. Das ist zu einem selbstverständlichen Teil des jungen und urbanen Lebensstils geworden. Wie das Automobil. SEAT bringt beides zusammen, mit komfortablen und sicheren Connectivity-Lösungen.

EN   Life is always a mixture of tension and relaxation, of action and calm. In a
SEAT, you can feel the pulse of the world – as long as you want to. But a SEAT is
also a private space for feeling at ease.

ES   La vida es siempre una mezcla de tensión y relax, de acción y de calma, según el
momento. Por eso, en un SEAT se puede sentir el latido del mundo, pero también es
un espacio privado donde sentirse cómodo.

DE   Leben ist immer eine Mischung aus Anspannung und Entspannung, aus Action und
Ruhe. In einem SEAT fühlt man den Puls der Welt – solange man das möchte.
Ein SEAT ist aber auch privater Lebensraum, in dem man sich einfach wohlfühlt.

PRIVATE SPACE

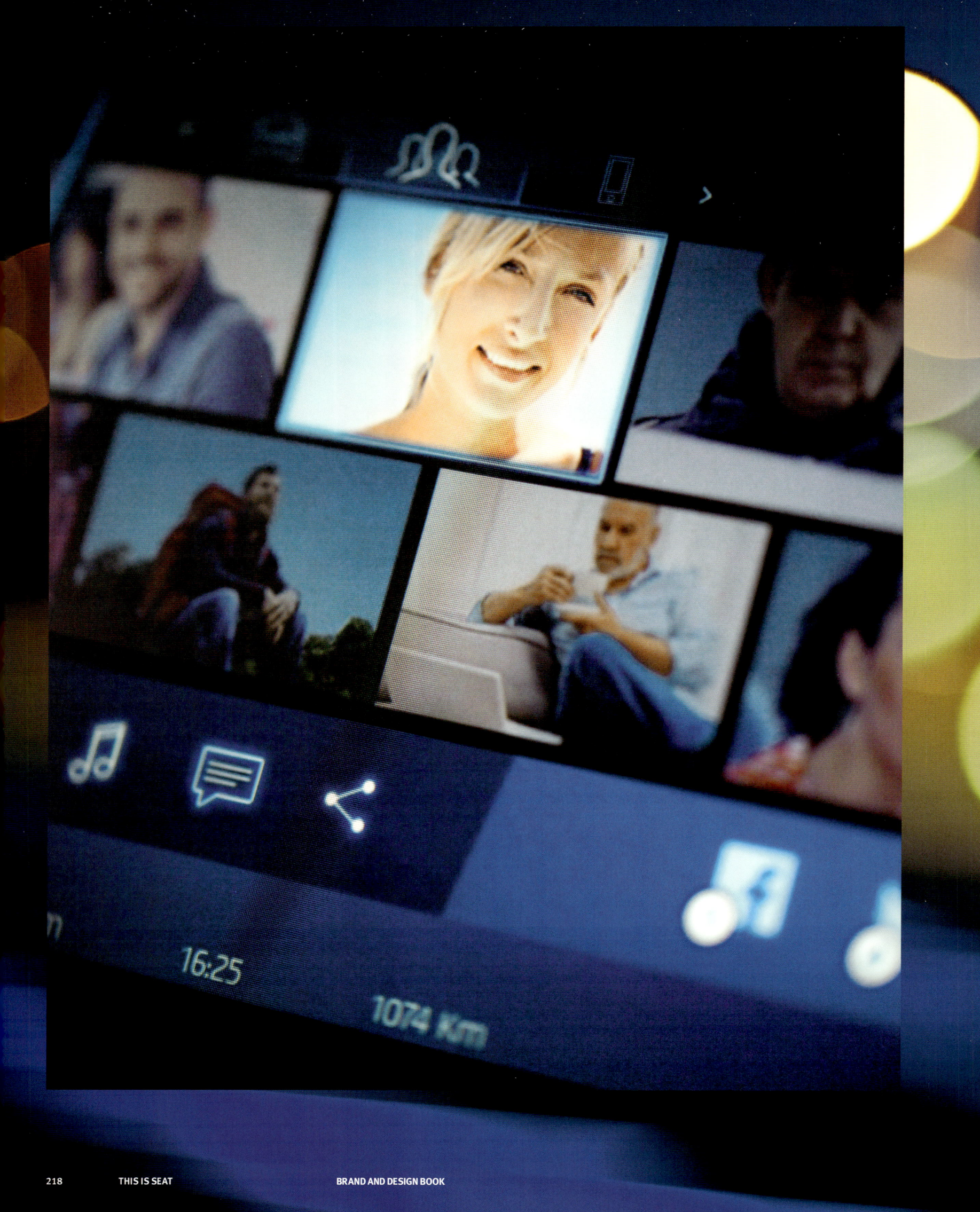

EN Share your life with friends – simply let them participate in your own experiences. It happens completely intuitively in a SEAT with connectivity powered by Samsung – while you remain fully focused on the road ahead.

ES Comparte la vida con tus amigos, hazles partícipes de tus experiencias. Esto resulta completamente intuitivo en un SEAT con conectividad de Samsung, siempre con la vista puesta en la carretera.

DE Sein Leben mit Freunden teilen. Sie einfach teilhaben lassen an den eigenen Eindrücken und Erlebnissen. In einem SEAT mit Connectivity powered by Samsung gelingt das völlig intuitiv – bei voller Konzentration auf die Straße.

SHARED LIFE

EN Life in motion – enjoy individual freedom. Try out new ideas, in a new car – without limitations – because SEAT is as multi-faceted as life itself.

ES La vida en movimiento. Disfruta de tu libertad, atrévete a probar cosas nuevas en un coche que no tiene límites, porque SEAT es polifacético como la vida misma.

DE Leben in Bewegung, die individuelle Freiheit genießen. Neue Ideen ausprobieren, in einem neuen Automobil. Ohne Einschränkungen. Denn SEAT ist so vielseitig wie das Leben.

MOTION

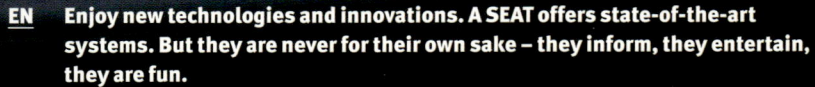

EN Enjoy new technologies and innovations. A SEAT offers state-of-the-art systems. But they are never for their own sake – they inform, they entertain, they are fun.

ES Disfruta de las nuevas tecnologías y la innovación. Un SEAT ofrece unos sistemas de última generación incluidos con cabeza: sirven para informar, entretener y divertirse.

DE Neue Technologien und Innovationen genießen. Ein SEAT bietet modernste Systeme. Aber sie sind nie Selbstzweck, sie informieren, sie unterhalten, sie machen Spaß.

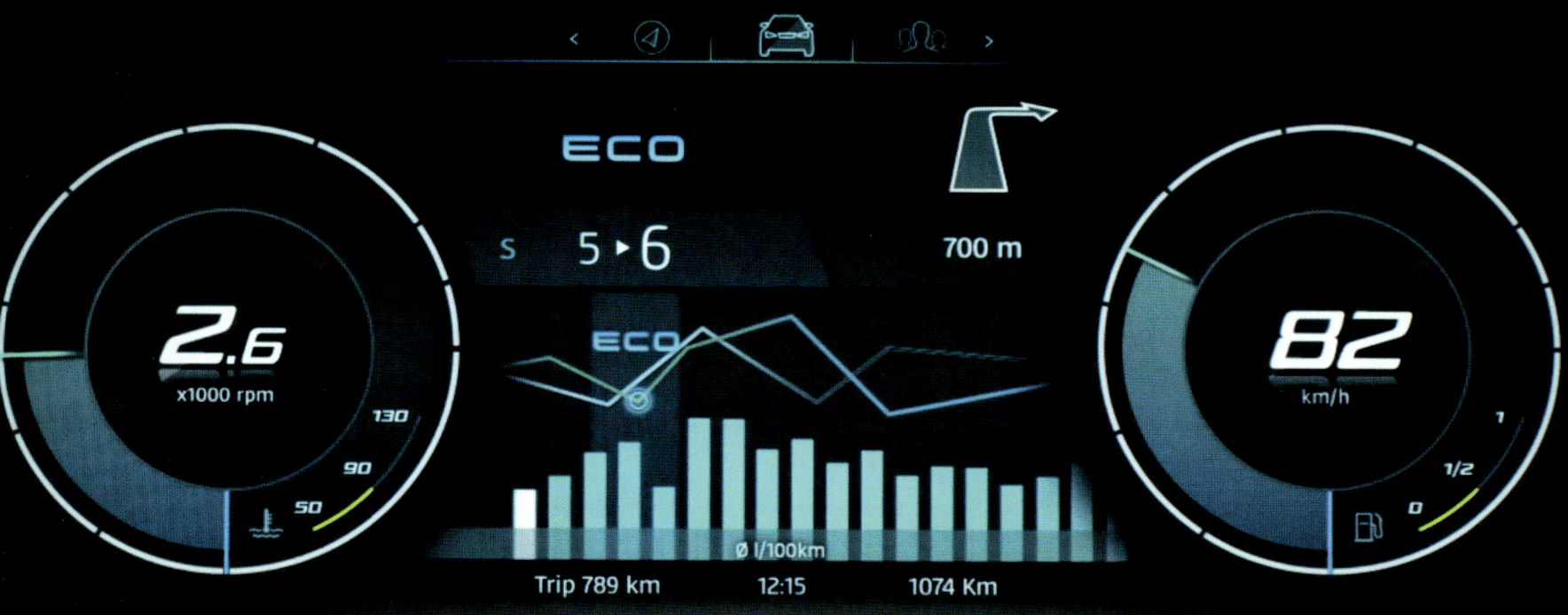

EN And rev it up once more – feel the performance, experience the sporty character with a car that is the perfect expression of your own approach to life.

ES Acelera, siente las prestaciones, experimenta el carácter deportivo de un coche que es la imagen perfecta de tu actitud ante la vida.

DE Und nochmal richtig aufdrehen. Die Dynamik spüren, Sportlichkeit erleben. Mit einem Automobil, das perfekt das eigene Lebensgefühl zum Ausdruck bringt.

ACTION SPIRIT

EN   Always the best of both worlds – passion and engineering skill, performance
     and safety, design and quality. SEAT – the perfect companion for a multi-faceted
     lifestyle.

ES   Condensar siempre lo mejor de ambos mundos: pasión e ingeniería, prestaciones
     y seguridad, diseño y calidad. SEAT es el compañero perfecto para un estilo de vida
     polifacético.

DE   Immer das Beste aus beiden Welten: Begeisterung und Ingenieurskunst.
     Dynamik und Sicherheit. Design und Qualität. SEAT – der perfekte Begleiter für
     einen vielseitigen Lebensstil.

SEAT

EN Because in the end everything we do is to make beautiful things
    happen – for You.
ES Porque, en definitiva, todo lo que hacemos es crear objetos de gran belleza
    para nuestros clientes.
DE Am Ende hat alles, was wir tun, nur ein Ziel: schöne Dinge möglich
    machen – für Sie.

EN Because in the end everything we do is to make beautiful things
    happen – for You.
ES Porque, en definitiva, todo lo que hacemos es crear objetos de gran belleza
    para nuestros clientes.

IMPRINT

THIS IS SEAT
SEAT Communications

Concept and text by reilmedia
Hermann Reil

Design by stapelberg&fritz
Davide Durante
Daniel Fritz
Roman Heinrich
Julian Hölzer
Maik Stapelberg

Illustrations by Barbara Stehle

Photographed by
Robin Wink Photography
David Breun
Jordi Gibert – Vistadiferent
Miquel Liso – Vistadiferent

Post production by
Wagnerchic Digital Artwork

Translation by
Elaine Catton (English)
MSS Barcelona (Spanish)

Text editing and proof reading by
Antonio Valdivieso
Fernando Salvador
Winfried Stürzl
Johannes Köbler

Production by Nele Jansen, teNeues Media
Color separation by Medien-Team Vreden

© 2015 teNeues Media GmbH + Co. KG, Kempen

Published by teNeues Publishing Group

teNeues Media GmbH + Co. KG
Am Selder 37, 47906 Kempen, Germany
Phone: +49-(0)2152-916-0
Fax: +49-(0)2152-916-111
e-mail: books@teneues.com

Press department: Andrea Rehn
Phone: +49-(0)2152-916-202
e-mail: arehn@teneues.com

teNeues Publishing Company
7 West 18th Street, New York, NY 10011, USA
Phone: +1-212-627-9090
Fax: +1-212-627-9511

teNeues Publishing UK Ltd.
12 Ferndene Road, London SE24 0AQ, UK
Phone: +44-(0)20-3542-8997

teNeues France S.A.R.L.
39, rue des Billets, 18250 Henrichemont, France
Phone: +33-(0)2-4826-9348
Fax: +33-(0)1-7072-3482

www.teneues.com

ISBN 978-3-8327-3300-1

Printed in the Czech Republic

Bibliographic information published by the Deutsche
Nationalbibliothek.

The Deutsche Nationalbibliothek lists this publication
in the Deutsche Nationalbibliografie;
detailed bibliographic data are available in the
Internet at
http://dnb.d-nb.de.

MIX
Papier aus verantwortungsvollen Quellen
Paper from responsible sources
FSC
www.fsc.org
FSC® C005833

teNeues Publishing Group
Kempen
Berlin
London
Munich
New York
Paris

teNeues